Edward Henry Sieveking

On Epilepsy and Epileptiform Seizures

Their causes, pathology, and treatment

Edward Henry Sieveking

On Epilepsy and Epileptiform Seizures
Their causes, pathology, and treatment

ISBN/EAN: 9783337418618

Printed in Europe, USA, Canada, Australia, Japan

Cover: Foto ©berggeist007 / pixelio.de

More available books at **www.hansebooks.com**

ON

EPILEPSY

AND

EPILEPTIFORM SEIZURES.

Their Causes, Pathology, and Treatment.

BY

EDWARD HENRY SIEVEKING, M.D.

FELLOW OF THE ROYAL COLLEGE OF PHYSICIANS;
PHYSICIAN TO, AND LECTURER UPON MATERIA MEDICA AT, ST. MARY'S HOSPITAL;
PHYSICIAN TO THE LATE DUKE OF CAMBRIDGE;
FELLOW OF THE ROYAL MEDICAL AND CHIRURGICAL SOCIETY;
CORRESPONDING MEMBER OF THE MEDICAL SOCIETIES OF HAMBURGH AND
CONSTANTINOPLE; ETC. ETC.

SECOND EDITION, REVISED AND ENLARGED.

LONDON:
JOHN CHURCHILL, NEW BURLINGTON STREET.
MDCCCLXI.

LONDON:
SAVILL AND EDWARDS, PRINTERS, CHANDOS STREET,
COVENT GARDEN.

TO HIS COLLEAGUES,

THE

PHYSICIANS AND SURGEONS OF ST. MARY'S HOSPITAL,

IN ACKNOWLEDGMENT OF MUCH KINDNESS,

AS A PROOF OF GOOD FELLOWSHIP,

AND

IN EVIDENCE OF HIS DESIRE TO EMULATE THE INTELLECTUAL

AND SOCIAL QUALITIES WHICH DISTINGUISH THEM,

THIS ESSAY

Is Dedicated

BY THE AUTHOR.

PREFACE TO THE SECOND EDITION.

The subject of epilepsy is one that embraces, perhaps more than any other limited department of medical science, the whole range of pathology and therapeutics. It is with this conviction that I have laboured at the changes which the present edition seemed to demand. I have sought in it to bring to bear not only my own matured experience, but also the advances in physiology and pathology with which the last few years have teemed.

The first edition was brought out a little more than three years ago, under circumstances which might palliate certain crudities of composition. As, however, it met with a reception beyond its merits, I indulge in the hope that its successor, worked out with the same love, though with more care, may not prove unacceptable to my medical brethren.

Although I have retained the mould in which the work was originally cast, I have, acting upon the advice of one well qualified to express an opinion, omitted the summary of cases given in the former edition. Instead of it I have interspersed numerous illustrative cases, detailed more at length, through the body of the work. I trust that I have shown sufficient consideration for the reader's patience in not admitting a larger number than was essential to the argument.

I have also appended a few of the formulæ of some of the medicines that I have alluded to in the body of the work, and which may, perhaps, be useful to the younger members of the profession who do me the honour to peruse the following pages.

. Throughout it has been my endeavour to be as exact as possible, and to use only such material, either of other authors or of my own, as belonged to the domain of fact. I have to the best of my ability avoided the introduction of hypothetical conclusions and vague statements; still no one can be more sensible of the shortcomings of the author than the author himself, but if my professional brethren confer upon the present volume that meed of praise which rewarded my first edition, my ambition will be fulfilled.

E. H. S.

17, MANCHESTER SQUARE,
March 30, 1861.

PREFACE TO THE FIRST EDITION.

THE strides that have been made during the last fifty years in the science of physiology can scarcely be said to have been equalled by the advances of pathology and therapeutics. This remark applies even more to the relation existing between the physiology and pathology of the nervous system than of any other class of organs or functions.

The labours of Bell, Marshall Hall, Flourens, Magendie, Müller, Brown-Séquard, have illumined a field which before the beginning of the present century was enveloped in darkness; and though the light thus shed upon medicine has removed much that was obscure, though Alison, Romberg, Todd, and Holland, Bright, Longet, Watson, Lebert, and many others, have helped to reconstruct the foundations of medicine upon the newly-acquired physiological basis, it cannot be denied that much of what is still taught in schools, regarding the treatment of disease, and of nervous maladies in particular, rests only upon empiricism.

But although the physician is compelled to admit that his therapeutic resources have not yet acquired the same firm scientific basis that he finds in some departments of medicine, he knows that all the laws regarding the preservation of health, the prevention of disease, the prolongation of life by these means, have been reduced almost to absolute certainty. Whether we take into consideration the diseases of one organ or system of organs, or another—whether we regard the diseases of childhood, manhood, or old age; in all cases the paramount importance of carrying into practice the rules of prevention embodied in sanitary science, holds good. Here lies our strength, here our logical consistency. No caviller or scorner can oust us from the strong position which here we occupy; for statistics, reason, and empiricism only serve to fortify it, while they mutually support one another.

While we concede a very high rank to sanitary science—a rank which may be said to bear the same relation to ordinary therapeutics that a multitude bears to the individuals composing it,—we may profit by the lessons it teaches us, by labouring for the prevention of disease wherever we meet with it in the daily execution of professional duties.

The sick man necessarily desires, above all things, to be freed from his present malady; but while the physician uses every appliance at his command to realize his patient's wishes, he must take a wider scope, and seek to prevent the return of the disease, or to anticipate where the seeds have been or may be sown, but have not yet, to the eye of the superficial observer, sprung into life.

In this direction, the study of diseases of the nervous system promises more fruit than we can expect

PREFACE TO THE FIRST EDITION.

to gather through the intervention of the Materia Medica alone. This, at least, is my hope and my conviction, seeing how little has been achieved in the treatment of some forms of nervous diseases, although they have been the subject of earnest study from the beginning of medical science.

In epilepsy especially the results have been, if not barren, yet unsatisfactory; and still the very fearful nature of the disorder, the strange and violent symptoms that characterize and almost seem to remove it from the domain of ordinary diseases, constantly attract new inquirers, each anxious that he may succeed in lifting the veil that shrouds the mystery. No individual, however, can expect to do this; but each may, in his sphere, seek to aid in its ultimate removal; and in this hope I have not shrunk from devoting much labour and time to a subject promising so little immediate reward.

The labour has been a labour of love; nor has it been undertaken without a full sense both of the difficulties in the way of the inquirer, and of the dangers of dogmatism and empiricism besetting his path.

With our present knowledge of hygiene, public and private, and with our acquaintance with the physiology of the nervous system, we are justified in expecting more decided benefit from a full development of our hygienic resources in combating epilepsy than by reiterated experiments with drugs alone. Although I have a high opinion of the value of medicines in the strict sense of the word, I place even a greater reliance upon the influence of hygienic and regiminal agents in combating chronic disease; and believe that no medical man avails himself sufficiently of the powers at his command who does not constantly combine both modes of warfare. Upon these principles my own practice is based, which, in theory at least, it is my constant endeavour to render as rational as possible.

Although I have repeatedly published papers in the journals on Epilepsy and allied subjects, the present work stands by itself. It is in no way a republication of what has already appeared; although, as a matter of course, I have used the same cases as the basis of the results to which I shall direct the reader's attention. These observations have been tabulated, and are given in an abridged form at the conclusion of the book. With the exception of cases of too recent a date, I have put together all of which I have preserved notes, and which have been under my own care.

Of the results of my inquiries I will here say nothing, except to express my vivid sense of the fact that my achievements fall far short of the goal to be attained. I may, however, add a hope that, although my readers may differ from some of the views which I entertain, they may feel willing to admit that, in my pursuit after light, I am not found to be running after a will-o'-the-wisp.

EDWARD H. SIEVEKING.

17, MANCHESTER SQUARE, LONDON,
September 9, 1857.

CONTENTS.

CHAPTER I.

Introductory remarks — General description of epilepsy — The phenomena of the paroxysm — The character of the convulsions and of the spasm — The description of Lucretius — The condition of the eye — The state of lungs and heart — Insensibility — The definitions given by classical writers — Author's remarks thereon . . . . pp. 1—13

CHAPTER II.

Importance of an early recognition of epilepsy — Analysis of the individual symptoms — The premonitions — The aura, either subjective or objective — It's frequency — The premonitory fear — Curious features in the aura — Frequency of the aura in the author's experience — Its varieties — The unconsciousness of epilepsy — The convulsions, clonic or tonic, occasionally absent — Exceptional phenomena — Marshall Hall's views
pp. 14—37

CHAPTER III.

Analysis f individual symptoms continued — Biting the tongue — The pulse in the paroxysm — Frequency of the fits — Periodicity — Lunar influences — Mead's case — Nocturnal and diurnal influences—Boyd's observations—Seasonal influences—Moreau's observations — Headache — Cases of epileptic headache — Delirium and hallucinations . . . . pp. 38—60.

CHAPTER IV.

The phenomena observed during the free intervals — State of chylopoietic viscera — Pupil — Vertigo — Petit mal — The sequelæ of epilepsy — Loss of memory — Fatuity — Paralysis — Moreau's statistics — Impaired circulation — Derangement of intellect — Fatality of epilepsy — Statistics of Registrar-General — Author's calculations . . . . pp. 61—84

CHAPTER V.

The cause of epilepsy — The demoniac controversy — Adams' opinion — Bucknill's views — Opinions of Churchmen — The frequency of epilepsy in England — In different troops — In the whole population — In France — Boudin's statistics — Epidemics of epilepsy — Sex — Age — Hereditary influences — State of the organs of excretion and secretion — The kidneys — The digestive organs — Costiveness — The sexual organs — Marriage — Masturbation — Sexual derangement in males and females — Should epileptics marry ? . . . . . . pp. 85—142

CHAPTER VI.

The exciting causes of epilepsy, primary and secondary—Psychical, physical influences — Diurnal changes — Smith's experiments —Syphilis—Sleep—Intemperance—Centric and eccentric epilepsy — Short's, Darwin's, and La Motte's cases.
pp. 143—160

CHAPTER VII.

The pathological anatomy of epilepsy—The import of the lesions found — Boyd's investigations — Ferrus' and Parchappe's inquiries—Wenzel's autopsies—Various lesions—Schroeder van der Kolk's observations — Remarks — Esquirol's inquiries— Further remarks . . . . pp. 161—188

CHAPTER VIII.

The theory of epilepsy—General remarks—Its proximate cause—Esquirol's classification—Prichard's view—General remarks—Sir A. Cooper's experiments—Schroeder van der Kolk's theory—Brown-Séquard's experiments and doctrine — Kussmaul and Tenner's views and experiments—Remarks—The influence of habit—Relation of epilepsy to kindred diseases—Metastasis in epilepsy — State of the blood in epilepsy — Handfield Jones' case and remarks—Acute and chronic epilepsy—Cases in illustration of author's remarks . . . . pp. 189—232

CHAPTER IX.

The treatment of epilepsy—During the paroxysm—Removal of all restraint—Avoid over-active treatment—Cool air—Sinapisms and the like—Compression of carotids—Cold applications—Galvanism—Volatile stimulants—Treatment of premonitory symptoms—Illustrative cases by the author—Ligatures—Dry-cupping—Internal remedies—Radical treatment of the disease—Trephining—Moral treatment—Derivation from the head—Various counter-irritants: the actual cautery, blisters, setons, tartrate of antimony ointment—The abstraction of blood—Illustrative cases—Ligature of carotids—Purgatives—Turpentine—Salts of iron—Case — Zinc — Cases — The valerianates — Silver — Frommann's case—Nitro-muriatic acid—Digitalis—Iodides — Bromides — Cases — Belladonna — Indigo — Cotyledon umbilicus—Mistletoe . . . . pp. 233—299

CHAPTER X.

The hygienic treatment of epilepsy—Influence of air—Exercise—Baths: special directions for their employment—Quantity and quality of food—Period for taking it—Rest of body and mind—Balance of the mental powers—Education of children—Their home management — Quack remedies, and their influence on the mind—Moral regimen—The physician a school-inspector—Conclusion . . . . pp. 300—321

Appendix of Formulæ . . . . pp. 322—328 .

ON EPILEPSY

AND

EPILEPTIFORM SEIZURES.

CHAPTER I.

Introductory remarks — General description of epilepsy — The phenomena of the paroxysm — The character of the convulsions and of the spasm — The description of Lucretius — The condition of the eye — The state of lungs and heart — Insensibility — The definitions given by classical writers — Author's remarks thereon.

THE history of epilepsy, more than of other affections of the nervous system, until the most recent periods has been the history of one of the weakest sides of medical science; it exhibits the inadequacy of our therapeutic measures to cope with a mysterious power in a painful degree. The prominent features of the disease were too palpable not to be early recognised; while their violence and suddenness struck the mind of the laity with awe, inducing in the beholder a conviction of an influence altogether removed from the ordinary phenomena of disease. The prevalent belief in the demoniac causation of epilepsy was combated by Hippocrates as it has been by recent writers of our own era; while even to the present day the ancient belief in lunar influences, as connected with the production of epilepsy, holds its ground in, as well as out of, the profession. Nor have there been wanting at all times earnest inquirers, who have carefully studied the disease in question; but if their investigations were barren, or, at least, fraught with less positive results than have rewarded students in other departments of pathology, this is mainly attributable to the fact, that the physiology of the nervous system has been, until the second quarter of the present century, a domain unknown and uncultivated by medical men. Without a definite physiological basis no derangement of function can possibly be understood; and although we have not yet attained to a solution of all the difficulties connected with the subject before us, there can be little doubt that we are in the right way to its achievement. Before entering into an analysis of the present theory or pathology of the disease, based upon recent advances of physiology, it will be well to examine its phenomena in detail.

The features that characterize epilepsy are symptoms of a powerfully convulsive character, sudden and brief; which, while they undoubtedly affect the general health of the sufferer, and induce him to attribute his entire morbid condition to them, also attract the attention of his friends and of the medical man generally so largely as to cause less prominent symptoms to be overlooked, which, however, belong as much to the entire portrait of the disease as the paroxysm. It is not to be wondered at that the suddenness of the seizure, the violence of the accompanying phenomena, the apparent inability of art to contribute much to their alleviation or arrest, should so often paralyze the hopes of the patient and of his friends. If we turn for an instant to other forms of disease, we shall find analogous instances, in which morbid conditions prevail, which do not attract

PRELIMINARY REMARKS.

much attention or cause serious apprehension until a climax is reached which taxes the entire vigour of the system, or proves beyond the control of nature or of art. Thus, during the prevalence of cholera the noxious influence of the poison is manifested very generally throughout the population by the almost universal occurrence of diarrhœa, which is amenable to simple hygienic and medicinal treatment, and is undoubtedly due to the same poison which induces the more fatal form to which the name of cholera is applied.* It has been shown that, while the curability of algide cholera by art is altogether a matter of doubt, there is no doubt that we may prevent or anticipate it by arresting the mild form which it puts on in the premonitory diarrhœa. We shall see that something similar occurs in epilepsy; and that our knowledge of the disease will be warped and very imperfect, so long as we have regard only to the fit that embraces all the characteristics to be found in scientific definitions, and overlook those minor symptoms which occur during the free intervals, but do not amount to the sad dignity of a paroxysm. It will, however, be convenient to commence the observations I have to make upon the disease by a delineation of a well-marked paroxysm or seizure itself. Having done this, I shall enter upon the bearing and value of the individual elements or symptoms of the paroxysm, and then examine the state of the patient during what is termed the free interval.

A brief and peculiar sensation, which has received

---

* See Report of the Committee for Scientific Inquiries in relation to the Cholera Epidemic of 1854, p. 10. Untersuchungen und Beobachtungen über die Verbreitungs-art der Cholera, &c. Von Dr. Max Pettenkofer. München, 1855.

the technical denomination of an *aura*, and which varies much in character, announces in many cases to the person on the point of being attacked that something peculiar is about to happen. The patient is suddenly deprived of consciousness and sensibility, a deadly pallor overspreads the face, and, often uttering a scream of a shrill, unearthly character, he falls as if hit by a gunshot, striking against anything that may intercept him, and injuring himself more or less, according to the nature of the opposing obstacle. There is entire unconsciousness; hence the patient does not protect himself, and falls into the fire, or into the water, upon sharp angles, or upon the rug, without any regard to personal consequences. It is this that renders extreme watchfulness necessary on the part of the attendants of epileptic subjects, while it suggests a ready means of recognising the malingerer or impostor. One of my patients, a boy of eight years, was nearly drowned by falling into a canal; another, a man, dislocated his jawbone in a fit; a third again, a female, broke her collarbone. Such cases might be multiplied, and minor injuries, cuts, bruises, and the like, are innumerable.

The pallor of the countenance yields to a livid flush, deepening with the increasing convulsions that supervene; the neck often swells, the surface veins of the head and neck in these cases becoming distended, as if about to burst; the tongue is thrust out between the teeth; respiration is laboured, and often suffocative; and bloody froth issues from the mouth.

This order of the phenomena, however, may vary or be modified in every possible form. I have in the same individual seen intense lividity and jactitation in one attack, who in another showed nothing approaching to venous congestion, where, in fact, one would have rather been disposed, from the throbbing of the arteries and the high complexion, to assume active determination of blood.

The limbs and features are convulsed and distorted, and it seems as though the restraining power of the muscular system had been destroyed or taken captive, and were running riot under some hidden but uncontrollable influence.

There is something necessarily suggestive of the influence of demoniac agency in these spasmodic throes; and without entering into a theological discussion on the terminology employed by the ancients or in sacred writ, it may be admitted that the epileptic paroxysm is peculiarly of a character to convey the impression that it is altogether beyond the ordinary range of physiological disturbances.

The convulsions that ensue are more or less violent, and generally show a predominance on one side of the body. The extremities are thrown about in sudden jerks, much like the movements of an animal recently dead whom we submit to the shocks of a galvanic battery. It has appeared to me that the left side is the one most frequently affected. The spasmodic action is shown most in the flexors, in some of which it amounts to a permanent tonic contraction; the thumbs are drawn across the palm of the hand, and the fingers are at times clenched so firmly, that when relaxation takes place the nails are found to have left deep marks. Similarly the toes are forcibly flexed. The closed fingers are popularly regarded as so essential an element in the disease, that it is commonly considered necessary to force the thumbs and fingers open; a proceeding which is utterly useless, and can lead to no beneficial result.

The spasm is shown very uniformly in the muscles that close the jaw. The spasm of the muscles of the jaw is, like the spasm of other parts, clonic or tonic. In the former case the result is that the patient grinds his teeth; and this is often done with such force as to break them. In the latter, the mouth is so firmly closed, that it may be impossible to open it or to introduce anything. This spasm is sometimes sufficiently violent to cause dislocation of the jaw. Dr. Cooke* quotes a case from Van Swieten in which this occurred. In consequence of the spasm, the inside of the cheeks and the tongue are very frequently lacerated and bitten by the patient. Hence bleeding at the mouth is a common symptom of the epileptic paroxysm, and one that may be relied upon as a pathognomonic sign, since persons who feign the disease will not be aware of its importance as a symptom, nor have the courage to produce a lesion sufficient for the purpose. If, as is frequently the case, the tongue is thrust out of the mouth during an interval of relaxation of the masseter and pterygoid muscles, and the jaw then spasmodically closed upon it, the protruded part becomes livid from congestion, and the appearance of the patient is rendered more than ordinarily frightful. The tongue, or rather a portion of the organ, has actually been bitten off during the paroxysm; but this is a very rare occurrence, since the spasm is generally relaxed before the teeth are able to sever the protruded part.

Owing to the saliva accumulating in the mouth, and the breath being entangled in it, the patient in expiration forces out froth from between his lips, and the froth is

---

* Treatise on Nervous Diseases, vol. ii.

frequently bloody if any abrasion or lesion has occurred within the mouth,

Ut fulminis ictu
Concidit et spumas agit, ingemit et tremit artus,
Desipit, extentat nervos, torquetur, anhelat
Inconstanter et in jactando membra fatigat.*

The spasmodic action is also shown in the voluntary muscles of the eyes, which are frequently turned up so as to leave little more than the sclerotic visible, the pupil and iris being concealed under the upper lid. This is not, however, a uniform or persistent symptom, as the eye may be closed, or the upturning only be brief. The pupil is commonly contracted, but by no means, as would appear from the statements of authors, invariably so. I have seen them very much dilated. I have observed them vary at different periods of the same paroxysm. Thus I have found them dilated at the early part of the attack, and subsequently become contracted; but at both times from internal causes, the influence or withdrawal of light not altering the condition. They are, however, always insusceptible during the profound paroxysm to the stimulus of light, and this fact may be used as a means of diagnosis in cases

---

* T. Lucretii Cari, De Rerum Naturâ, libri sex. Lib. iii. v. 486. Lucretius in these lines and those subsequently quoted clearly describes an attack of epilepsy; he uses it as an illustration of his argument regarding the immortality, or rather the mortality of the mind (*animus*); he is of opinion that because the immaterial principle is affected by disease, and is liable to be cured by remedies that cure the body, like the latter it is destructible:—

Et quoniam mentem sanari, corpus ut ægrum
Cernimus et flecti medicina posse videmus,
Id quoque præsagit mortalem vivere mentem.

of feigned epilepsy. It is manifest that the iris partakes in the spasmodic action affecting the muscles generally, and this may be either tonic or clonic. The congestion that takes place to the head is particularly shown in the eye becoming suffused and bloodshot, at times leaving considerable ecchymoses in the conjunctiva.

Small extravasations have been seen under the skin of the head and face from the same cause. Internal congestion is also found to affect the brain and meninges, as is demonstrated in those cases where epilepsy proves fatal in the paroxysm itself.

The breathing is short and hurried, as in a person violently agitated; the disturbance of the respiration appears to be proportionate to the violence of the paroxysm. The same applies to the action of the heart; but no derangement of either function is observed to be connected with the epileptic paroxysm which can be regarded as in any way pathognomonic.

In severe and prolonged seizures, either from exhaustion or from being poisoned by the highly carbonized blood, the heart appears to become so oppressed as almost to cease contracting. In one case of deep interest to myself, a fatal issue seemed actually to have taken place, from the pulse getting gradually feebler and appearing altogether to stop; the patient, however, rallied and survived for some years.

The abdominal viscera do not suffer in any characteristic way. In violent seizures, the contents of the stomach are at times expelled by vomiting; occasionally involuntary defecation and micturition occur during or towards the close of the fit, and erections and seminal discharges have also been observed to take place. None of these symptoms are sufficiently uniform to lay claim to the dignity of a pathognomonic sign. They are, however, of importance when they occur, as indicating the extent to which the system is under the control of the disease.

During the continuance of the symptoms of which I have hitherto spoken, the insensibility of the patient continues complete; we are unable to rouse him by any of the means that would suffice to awake even a profound sleeper; the mental functions are in complete abeyance, and even the reflex actions are arrested. After the lapse of from ten to twenty minutes the convulsive movements subside, profuse perspiration commonly ensues, consciousness returns, and the patient is restored to comparative health. A temporary swelling of the face is sometimes observed after the fits; after repeated attacks, a bloated condition often remains, which gives the confirmed epileptic a peculiar recognisable expression.

> Ubi jam morbi reflexit causa, reditque
> In latebras acer corrupti corporis humor,
> Tum quasi vaccillans primum consurgit, et omnis
> Paulatim redit in sensus animamque receptat.*

The fit is very commonly followed by considerable drowsiness; the patient sinks into a deep sleep, sometimes almost comatose, but ordinarily presenting no morbid features; it appears to be granted by nature for the purpose of restoring the enfeebled and exhausted powers. At times a condition identical with that of delirium tremens persists for a day or two after the paroxysm. Headache, for a longer or shorter period, very commonly remains after the epileptic seizure. In mild cases, however, the patient recovers his ordinary state of health immediately, and feels as if nothing had happened.

---

* Lucretius, ibid.

Such is the ordinary course of an epileptic paroxysm; but though it is common to meet with the symptoms just detailed, they all, more or less, differ in frequency and uniformity.

The definition which Sauvages* gave of epilepsy was this: "Est morbus clonicus universalis chronicus et periodicus, cum sensuum feriatione in paroxysmo et ante-actorum oblivione." The description which we find in Galen runs thus: "Verum et epilepsia† con-

---

* Nosologia Methodica sistens Morborum Classes, &c. Auct. Franc. Boiss. de Sauvages, Regis consiliario, &c., tom. i. p. 578. Ed. ultima. Amstelod. 1768.

† Though "epilepsia" is here used by the learned translator of Galen, the more common appellation of epilepsy, employed by Latin authors, was *morbus comitialis*, which was derived from the fact that, if a person was seized with a fit during the *comitia*, the meeting was dissolved on account of the unfavourable omen supposed to be involved in the occurrence. Facciolati under *Comitialis Morbus* says, "Ratio appellationis est quia si quis forte ipso comitiorum tempore illo (sc. morbo) corriperetur, comitia dissolvi ac in alium diem differri necesse erat; ut ex *Festo* in *Prohibere* et ex *Seren.* Samm. discimus." It appears that Pliny, who speaks frequently of epilepsy, uses *comitialis* as a substantive for an epileptic (*e.g.*, Nat. Hist. lib. xx. cap. 5). In discussing the use of Staphylinos or Pastinaca erratica, he observes: "Philistio [doubtless an apothecary of the day], in lacte coquit at ad stranguriam dat radicis uncias quatuor; ex aqua hydropicis, similiter et opisthotonicis et pleuriticis et comitialibus."

I may take this opportunity of adverting to the large knowledge which Galen had already acquired of the disease, and to the fact that much of what we now teach respecting it was taught, as well as many of the errors which we still combat were recognised and combated by this great physician. Any one who desires only a superficial view of his teachings regarding epilepsy, will find instruction in perusing the admirable index to Kühn's edition of his collected works published at Leipsic in 1833. Tom. xx.

vulsio est omnium corporis partium, non continua ut emprosthotonos, opisthotonos, et tetanos sed quæ ex temporum accidit intervallis, neque solum hâc re sed mentis quoque ac sensuum oblæsione, a jam dictis convulsionibus differt; unde constat," he continues, "in superioribus corporis partibus, cerebro videlicet, hujus affectus ortum esse." Except that the second definition is rather more lengthy, the sense is essentially the same as that conveyed by the first.

And if we compare the description of more recent authors of epilepsy with the above two, we shall find but little altered. The prominent features of the paroxysms are, in fact, so marked, that there could be little difference in the accounts given of them. This opinion has been so fully realized by some, that, for instance, Dr. Lysons,* a physician who wrote upon epilepsy in the last century, prefaces his observations and cases with the remark that "the epilepsy is too well known to need description."

I quote one definition from a recent author, in order to show that the lapse of ages has not materially affected the views that prevail on the subject. Dr. Copland† defines epilepsy as "Sudden loss of sensation and consciousness, with spasmodic contraction of the voluntary muscles, quickly passing into violent convulsive distortions, attended and followed by sopor, recurring in paroxysms more or less regular." A better definition of the epileptic paroxysm, and one that more completely embraces its essential features, can probably not be devised; yet I incline to think

---

* Practical Essays upon Intermitting Fevers, &c. By Daniel Lysons, M.D. Bath, 1772.
† A Dictionary of Practical Medicine, vol. i. p. 785.

that our knowledge of the disease will not gain by definitions, however accurate, as they tend to limit the attention to the paroxysm, and often mislead the observer by inducing him to overlook conditions to which the definition does not apply, but which will appear, after a fuller examination of the whole subject, to belong to the same category of diseased conditions as epilepsy.

We should regard the fit as the flower of a noxious weed. We may be able to recognise the flower by its colour or its smell, when we have once seen it; but we shall fail to prevent the development of other flowers and seeds on the same plant, and on other plants, unless we recognise the plant itself by all its characters, and are thus enabled to pluck out the whole weed by the roots wherever it may be discovered.

I am far from indulging in the presumptuous opinion that I have succeeded in discovering the means of doing this; but it does appear important to establish the correct principle upon which to proceed. And I cannot but think that the difficulties which environ the whole question of the nature and treatment of epilepsy are enhanced by the paramount attention which has been paid to the fit itself.

These views are borne out by the observation of Dr. Watson, who objects to attempting a definition of epilepsy, on the ground that its forms are so various, and its modifications so numerous, that no general description of it can be given.*

Still, although it may not be possible to contrive a definition which shall embrace all the features observed

---

* Lectures on the Principles and Practice of Physic, vol. i. p. 609, seqq.

in epilepsy, it is essential that the medical man should be well acquainted with the individual phenomena uniformly or exceptionally accompanying the paroxysms.

In well-marked cases, the diagnosis offers no difficulties; but instances are often presented to us in which a correct judgment and interpretation cannot be arrived at without mature deliberation. An error may entail undue anxiety or alarm, or, on the other hand, prevent the adoption of prompt and energetic measures calculated to arrest the progress of the malady.

I shall therefore in the following pages proceed more minutely to analyze the different symptoms, with a view to establishing their nosological and diagnostic value.

CHAPTER II.

Importance of an early recognition of epilepsy — Analysis of the individual symptoms — The premonitions — The aura, either subjective or objective — Its frequency — The premonitory fear — Curious features in the aura — Frequency of the aura in the author's experience — Its varieties — The unconsciousness of epilepsy — The convulsions, clonic or tonic, occasionally absent — Exceptional phenomena — Dr. Marshall Hall's views.

THERE are convulsive diseases resembling epilepsy which must not be confounded with it; while syncope, hysterical spasms, apoplectic conditions, and the like, may also be mistaken for this affection. On the other hand, it frequently happens that an individual has attacks of an epileptic character which are not recognised until long after their first manifestation. I speak not only of spasmodic movements, subsequently proved to be identical with the aura of a complete paroxysm, of epileptic vertigo or headache, of temporary attacks of unconsciousness, and similar apparently trifling symptoms; but of actual epileptic seizures, affecting individuals at night and during sleep, under circumstances precluding them from being watched and from becoming themselves aware of their malady.

It has been previously stated that the different features which we have attributed to the epileptic seizure may vary in frequency and uniformity. It is manifestly important in point of diagnosis and treatment, as well as in reference to the views we may form of the intimate nature of the malady, to determine as nearly as may be their relative value. It will therefore be my endeavour to analyze them with all care, and to bring to bear upon their consideration as much of positive data as the history of medicine or my own experience supplies.

The first question that it may be proper to answer refers to the frequency of the occurrence of premonitory symptoms. The affection that indicates the approach of a paroxysm has, from the time of Galen* downwards, been denominated an aura; and it appears to be the prevailing doctrine among many physicians even to the present day, that the premonitory symptoms really occur in the form of an aura, or a breath of air. Esquirol,† in speaking of the aura, maintains the literal interpretation of the term, and says that the sensations spoken of by the patients : " Se propagent comme une *vapeur* le long des membres, du tronc, du cou vers la tête, et lorsque cette vapeur est arrivée au cerveau l'accès éclate." Whether Esquirol merely

---

* De Locis Affect., lib. iii. c. 2. In relating a consultation about an epileptic boy, Galen says that one of the physicians (ἄριστοι ἰατροί), in describing the premonitory symptom, called it an aura: αὔραν τινὰ ψυχρὰν ἔφασκεν εἶναι.

It is curious how authority has been misquoted with reference to this matter. It does not appear that Galen or the elder physicians implied that "the puff of wind" so commonly asserted to be premonitory of epilepsy really was an ordinary symptom. Still I have found no other passage than the one just quoted to which this opinion could be legitimately referred, and there it is manifestly merely the statement of an individual patient describing his own sensations.

† Des Maladies Mentales, tom. i. p. 274, seqq. Paris, 1838.

meant to convey that the symptom might be likened to a breeze, from the manner in which it was often found to pass from the extremities to the head, or whether his description is intended to imply that the patients felt a breeze or aura previous to the attack, I cannot decide. Certain it is that many of those physicians who have paid special attention to the phenomena accompanying epilepsy have failed to recognise the presence of an aura in the literal sense in the statements of their patients. Prichard* distinctly avers that he has never met with a patient who described the premonitory symptoms as an aura. By the list of the signs indicating the approach of a paroxysm in my own patients, it will appear that none of them described his feelings as a puff or draught.

Georget (quoted by Dr. Watson) states that he found premonitory symptoms not oftener than four or five times in one hundred cases.

My own experience tallies with that of Romberg, who says that he has met with premonitory symptoms in about one half of his cases of epilepsy.† Of one hundred and four cases of epilepsy,‡ which have been

---

* A Treatise on Diseases of the Nervous System, part i. pp. 85-113. London, 1822.

† A Manual of the Nervous Diseases of Man. By M. H. Romberg, M.D. Edited by E. H. Sieveking, M.D. Vol. ii. p. 197, seqq. Syd. Soc. Ed.

‡ I may take this opportunity for stating that wherever I refer to my own observations and analyses for statistics, I shall quote the conclusions given in two papers read before the Royal Medical and Chirurgical Society in 1857 and 1861, each based upon a careful analysis of fifty-two cases of epilepsy. Although they do not embrace all the cases of the disease which have fallen under my notice, the latter have been taken in the chronological order in which they presented themselves, and

under my own care, and of which I have preserved careful notes, forty-eight, or a little more than forty-six per cent., showed some indication of the approaching paroxysm. It must not, however, be concluded that because a patient at one time is made aware of the event about to take place, that therefore it will always be so. This Protean disease varies in this as in many other features; still it is most commonly the case that a patient habitually experiences a premonitory symptom, or that he is uniformly seized without any indication whatever.

The subjective sensations which the patients describe as preceding the fit are extremely various. But even after hearing the details of a small number, it cannot fail to suggest itself that they may, without an effort, be ranged in two classes; those that are referred to the trunk and extremities, and those that appear at once to affect the head. In the former case the sensation is always described as mounting towards the head, and in the majority of cases the paroxysm appears to strike down the patient on its reaching that part: in the latter the sensation commonly takes the form of some strange illusion, which, however, the patient is able to recognise as such.

Those who like to exercise their ingenuity in classification might also propose a division of premonitory symptoms according as they are more palpably physical or more of a mental or emotional character. It appears, however, on the whole, simpler to regard the latter as manifestations of cerebral derangement evidently depending on a physical cause.

---

without selection. I may have occasion to quote from my notes other cases which I have observed, but which from various reasons it was inconvenient to add to the above.

18   EPILEPSY AND EPILEPTIFORM SEIZURES.

Tissot, whose works may yet be consulted as models of close observation and clear reasoning, quotes from Peiroux* the case of a young man who, when his fits came on, thought he saw a carriage drive up at a gallop and with great noise, containing a little man in a red bonnet; fearing to be *écrasé* by the carriage, he fell down stiff and without consciousness. Dr. Watson relates that "the late Dr. Gregory, of Edinburgh, was assured by a patient of undoubted veracity, that always when he had a fit of epilepsy approaching, he fancied that he saw a little old woman in a red cloak, who came up to him, and struck him a blow on the head, and then he immediately lost all recollection and fell down."† In Tissot's work we find that even in sleep, during which epilepsy frequently supervenes, peculiar dreams may indicate the approaching paroxysm. He gives the case of a man who dreamt that he was pursued by a bull, and soon after waking was seized with a fit.

These are, however, rather the curiosities of epilepsy; the sensations of the patient not generally acting upon the sensorium in such a way as to produce illusions of the fantastic kind just described. With this exception, we may say that there is scarcely an impression referrible to the nerves of common or muscular sense, or of the special senses, which does not occasionally indicate the approach of an epileptic fit.‡ The premonitory symptom is generally accompanied by a

---

* Observat. Medicin., p. 90; in Œuvres de Monsieur Tissot. Nouvelle édition, augmentée et imprimée sous ses yeux, tome septième. Lausanne, 1790.

† Lectures, vol. i. p. 616. 1843.

‡ A very complete list may be found in Dr. Copland's Dictionary of Medicine, under the head of Epilepsy.

sense of fear and terror.* One of my patients described the sensation, which in him passed from the stomach to the head, of a pleasing character. Children particularly show the alarm they experience by running to and clinging to their nurses or mothers. The aura may be an undefined sense of indisposition or discomfort; it may be a definite pain, giddiness, or suffocating feeling; or it assumes the more classical form described as an aura, which is characterized by the passage of a peculiar sensation from some part of the body to the throat or head. In the case of the last we would specially observe that authors commonly state that when the aura, or sensation, reaches the head, the insensibility ensues; it has rather appeared to us that the patients refer the termination to the throat. With some patients the premonitory symptoms assume a more tangible form, and one that makes itself perceptible to bystanders.

---

* Moreau (Psychologie Morbide, Paris, 1859, p. 281) describes this common symptom with his usual felicity, and remarks upon it as follows:—" Parmi les signes prodromiques de l'épilepsie il en est un que les auteurs ont généralement noté, mais sans remarquer ce qu'il y avait de singulier au point de vue psychologique; c'est le sentiment de la peur. Tout à coup, au milieu de la santé physique et morale la plus florissante, dans le calme le plus absolu de l'esprit et du cœur, au sein des occupations les plus douces et les plus exemptes d'émotions, le malade se prend à *avoir peur*. Peur de quoi? Il n'en sait absolument rien, mais il a peur; mais il ressent, moralement et physiquement, tous les effets de cette passion; il frissonne de la tête aux pieds, son cœur bat, sa poitrine est oppressée, sa vue se trouble, ses yeux sont hagards, il s'écrie d'une voix étouffée, *J'ai peur! j'ai peur!*

"Ces phénomènes sont à rapprocher de cette explosion d'hilarité, de ce rire inextinguible qui s'empare de certains individus, menacé d'un accés de manie, ou bien encore des personnes qui ont pris du haschish."

Dr. Cooke* relates a case in which the approach of a paroxysm was indicated by a peculiar blue colour of the lips. "I. Frank," as quoted by Dr. Copland, "saw the paroxysm preceded by an eruption over the whole body, except the face, of the vitiligo alba." The same author states "that in twenty-one epileptics treated in the clinical wards of the hospital at Wilna, vomiting announced the paroxysm in seven." Symptoms that may be termed objective have presented themselves to me in the form of tremors, cough, sickness, rigors, a shaking of one hand, smacking of the lips, and spasms of other voluntary muscles.

Schenck relates a case of epilepsy which came under his own observation, in which the patient, before the seizure, was repeatedly turned round in a circle, and then fell to the ground in an ordinary paroxysm, "magna astantium commiseratione." Peiroux (quoted by Tissot) mentions a man who, before becoming unconscious, was compelled to run backwards ten steps; the unconsciousness was very brief, and he at once rose up again as if nothing had occurred. In Schenck we also find the account of a man, aged thirty, of whom it is said, in rather quaint Latin, "Solebat, quum duos vel tres passus progressus esset, sese inflectere quasi in circulum, idque continenter facere compulsus erat." This patient subsequently became epileptic, and the peculiar movements then ceased.† Such cases as those

---

* A Treatise of Nervous Diseases, &c., vol. ii. part 2. London, 1823.

† Παρατηρησεων sive Observationum medicarum, rararum, novarum, admirabilium, monstrosarum, Volumen. Studio atque opera Johannis Schenckii à Grafenberg. Francofurti, 1609, p. 110, where both the above cases may be found.

related by Schenck and Peiroux have received the name of "epilepsia cursiva," under which term Dr. Andree* details two well-marked instances, which were both cured by venesection, antiphlogistic remedies, and antispasmodics. They are instructive and well told, so as to justify our inserting one of them briefly here: "Rebecca Cole, ætatis sixteen. Before her seizures she first perceives a weight in her head, which makes her hang it down; then a tremor all over ensues and a sense of faintness; she then runs till she meets with some resistance, then falls down, struggles at first, after which she lies still and gradually recovers. The fit being over she trembles, is faint, sick at stomach, and dizzy; and now, by frequent returns of them, is almost become stupid."

Dr. Paget,† of Cambridge, has recently published a very interesting case in which frequent bursts of unmeaning laughter showed themselves in an epileptic subject as the precursor of his fits. They occurred day after day, and several times in the day; by day as well as by night when the patient was asleep. The laughter lasted about a minute, and then generally ended like the laugh of a person tickled by something ludicrous. It was not loud and unnatural like an hysterical laugh, and his face had the natural expression. There were other symptoms in this case which make it a curiosity in medical literature.

As I have not met with a statement in writers on

---

* Cases of Epilepsy, Hysteric Fits, and St. Vitus's Dance. By John Andree, M.D. London, 1746.

† A Case of Epilepsy with some Uncommon Symptoms, and a Commentary thereon. *British Medical Journal,* Sept. 22nd, 1860.

the subject, indicating the relative frequency of the various forms of premonitory symptoms, the following list of the warnings experienced by my own patients, and generally given in their own words, may not be devoid of interest.

1. Sense of choking and dimness. 2. A sensation extending from the thumb up the arm, with spasm of the latter. 3. Headache. 4. A sensation ascending from the stomach. 5. A sensation passing from the hand to the head. 6. Dimness and pain in the right arm; in this case the premonitory symptoms did not occur at the time the patient was under my observation, but had prevailed at an earlier period. 7. Pain across the shoulders. 8. Loss of sight. 9. Vertigo and general stiffness. 10. Feeling of illness for half an hour before the fits. 11. Head goes round. 12. Sense of suffocation and tremors. 13. A momentary warning. 14. Dimness. 15. Short cough. 16. Sickness. 17. Lightness of head, followed by oppression; in this case the premonitory symptoms were generally absent. 18. Sense of strangeness. 19. Pain at stomach and sickness. 20. Drowsiness. 21. Loss of power in left hand for twenty minutes before the fits. 22. Rigors. 23. Pain in hypogastrium. 24. Shaking and curious sensation in hand. 25. Lightness in head. 26. Sense of heavy weight. 27. Sometimes the fits preceded by a cry. 28. Pain in one or both temples. 29. Great fear. 30. Flushing for several hours before the attack, the latter being immediately preceded by a smacking of the lips. 31. Fear. 32. A sudden numbness of the head. 33. A peculiar sensation about the heart for five or six hours before. 34. Is cross and peevish for a day before the fits. I may remark apropos of this case that one of

the most curious features in epilepsy is the fact that in a great number of instances we find the patients in particularly good health and spirits at the time of their attacks; so much so, that some are almost afraid of feeling thoroughly well for fear of a seizure. 35. Formerly experienced a sense of dulness before the fits. 36. Drowsiness and general languor. 37. Lassitude and biliary derangement. 38. A sense of terror, with formication in the right hand, and passing upwards. 39. Sleeplessness and anxiety. 40. Some indefinite consciousness of the approach of a fit. 41. Spasm of the left arm and leg. 42. Creeping sensation from the tips of the fingers to the face, with spasms of the muscles of the parts through which the sensation passes. 43. A sensation mounting up the left arm to the head. 44. An indescribable idea. 45. Vertigo and flashes. 46. Turgidity of the face for two days before an attack. 47. A horrible sensation. 48. A feeling of something closing in the patient at times.

This list is sufficient to show that there is an entire absence of uniformity in the character of the symptoms; still we gather from it that the sensations referred to the trunk or extremities are more numerous than those which are described as having their seat in the head. In about seventeen instances the sensation may be classed as a cerebral phenomenon, while in the remainder its site is stated to be in other parts of the body.

In regard to diagnosis, the presence or absence of a premonition, as I have remarked in my second paper presented to the Medical and Chirurgical Society, offers no special indication; but it does so as to treatment, and it is therefore an important feature. It enables us to ward off an attack, and also allows the patient to place himself in a position where he shall not be likely to injure himself during the seizure.

The occurrence of the premonitory symptoms in the head or about the trunk and limbs respectively, is regarded by some as an indication of the character of the malady; the effective cause of the disease, the causa proxima, being supposed to reside in the head or elsewhere, according to the form of the "aura."

As this question involves the whole theory of the disease, I merely allude to its bearing upon the centric or eccentric origin of epilepsy, and pass on to the consideration of other matters.

Insensibility is ordinarily and very justly regarded as an essential symptom in genuine epilepsy.

In the complete paroxysm the unconsciousness is so profound that we are entirely unable to rouse the patient. The mind seems to be in abeyance. The patient on recovery has no recollection whatever of what has happened, and only learns by sad experience to explain the injuries which he may have received.

Even the reflex functions appear to be in abeyance, so that it has been suggested to introduce irritants, like snuff, into the nose, as a test of the reality of the seizure. Romberg doubts that this insensibility to reflex action is a uniform accompaniment of epilepsy; for he states that he has repeatedly observed the same start to follow the sprinkling of cold water over the face in the epileptic subject as in the healthy individual. "We have yet to inquire," he continues, "into the manifestations of reflex action resulting from irritation of the mucous membrane, as to whether, for instance, sneezing is produced by the introduction of acrid vapours into the nose, or cough by irritant fumes."

Though exceptionally there may be reflex action, it is necessary to bear in mind, and to impress upon the attendants of patients, the general fact of their entire insensibility, as great injury is sometimes done by the application of remedies calculated to rouse the sufferer, but which, on his recovery (which they have not influenced), are found to have induced local irritation, inflammation, and serious mischief.

Although the occurrence of entire unconsciousness is regarded almost universally as essential to constitute the epileptic fit, and enters into all the definitions, there can be no doubt of the existence of paroxysms which present all the other symptoms of epilepsy, but in which a certain amount of consciousness is retained throughout.

We shall have occasion to see that several of the diseases that are commonly regarded as residing mainly in the nervous system merge into one another, and that the boundaries by which they would appear to be circumscribed by nosologists are by no means so uniformly to be traced.

In the same way the symptoms belonging to epilepsy vary much. I have repeatedly been at a loss for a considerable time after a patient has come under my hands, to determine whether the affection he or she was labouring under was epileptic or not, until the occurrence of certain well-marked symptoms removed all doubt. I have had a patient under treatment, who presented all the symptoms of epilepsy, and had also bitten the inside of her mouth so as to cause hæmorrhage; she asserted that frequently in her fit she was aware of circumstances going on around, and that she retained sufficient memory of what has been spoken in her presence to repeat it after the paroxysm has subsided; at other times the unconsciousness was absolute. Nor is this the only case of the kind that has fallen under my notice. A gentleman who is under my care while writing these lines, informs me that he at times remembers having dreamt during the fit, showing that that complete abeyance of the mental faculties which ordinarily prevails does not always exist. Another, who in the severe paroxysms to which he is subject is profoundly comatose, states that in the minor attacks (*petit mal*) he has the sensation of seeing a large plain of snow.

Some authors of acknowledged reputation adopt the same view, that epilepsy may occur in which more or less consciousness is persistent throughout. It appears, in fact, difficult to understand how it can be reasonably denied by any one who admits that in the intervals of the true paroxysm, vertiginous and spasmodic attacks very commonly affect the epileptic patient, which evidently stand in a close relation to the main seizure, but are not characterized by loss of consciousness. A high authority in all matters connected with the pathology of the nervous system, Dr. Prichard,[*] who in everything that he has written evinces a mind capable of the largest views, remarks on the point in question, that in some cases there is a certain amount of consciousness throughout the fit, which he regards as being generally a prelude to the total abolition of the disease.

Poupart[†] relates the case of a lady in whom the fits were of such brief duration, that unless, as sometimes

---

[*] A Treatise on Diseases of the Nervous System, part i. London, 1822.

[†] Esquirol, Maladies Mentales; and, Mémoires de l'Académie Royale des Sciences, 1705.

happened, they were accompanied by a scream, persons in whose company she happened to be were not necessarily aware of anything being wrong. Her father had been epileptic, and she herself was seized at different times without being thrown down; the eyes were convulsed, the look became fixed for a few seconds, after which the patient, herself unconscious of the interruption, would resume the thread of her conversation where it had been dropped.

The following is a case of the kind, which but for the family history might perhaps be liable to a different interpretation: E. H., aged 32, the wife of a coachmaker, consulted me in 1856; she had lost her father from consumption and her mother by mammary cancer, and has two sisters who are epileptic. She had not herself ever had fits when she came under treatment, was of florid complexion and well made. Her catamenia were regular and normal, and she had one healthy child. For two years past she had suffered from occasional temporary loss of eyesight, with occasional loss of power, and numbness in her arms, and loss of speech. The attacks commenced in the hands and mounted up to the head (aura), and lasted about fifteen minutes; they were followed by severe headache, lasting two or three days, and occurred at intervals of from two to six weeks. The urine was normal, and contained neither sugar nor albumen. The left pupil was larger than the right, and she saw a little less well with the right than with the left eye. The pulse was 108, and soft; the tongue clean. She remained under treatment for above a month, at the end of which time the report is that she was much better, had less numbness, vertigo, and sinking, and that the functions of the body were well performed.

28 EPILEPSY AND EPILEPTIFORM SEIZURES.

This case acquires additional interest from the fact, that on inquiry I now find (March, 1861) that the patient whom I had entirely lost sight of two years ago, was seized with complete epilepsy, and has since been subject to it in an aggravated form.

I would also refer, in connexion with this question, to three important cases detailed by Dr. Bright,* in a paper on epilepsy from local disease, where the distinguished author dwells upon the fact of the patient retaining a certain degree of consciousness during the fits, which was regarded as an indication that they depended upon a local cause rather than upon a constitutional affection.

The convulsions which are associated with epilepsy present great variations in character and extent.† Generally they are clonic; and the violence with which the muscles act often renders it matter of physical difficulty to restrain the patients; whenever it entails no special risk to the patient, it is better that nothing be done to interfere by physical force. Moreover, the convulsions are extremely irregular; the want of coordination being a feature deserving special attention in our endeavour to localize the cause of the disease. The most ordinary movements that we meet with are an irregular jerking of the head to one side of the body, accompanied by contraction of the zygomatici and other facial muscles, and spasmodic action of the extremities of one side of the body. These movements may be incessant during the attack, or they may only occur at

---

* Guy's Hospital Reports, vol. i. p. 457.

† Nulla quippe gesticulatio, inflexio posituraque noscitur quam non aliquando exhibuerit epilepsia. — Boerhaave, Aphor. de cognosc. et curand. Morbis, p. 237. Lovanii, 1752.

longer or shorter intervals; or, again, they are so slight, or so limited in extent, as to be imperceptible during the greater part of the seizure. The duration and violence of the muscular spasm are at times so great as to induce a wonder that human nature can sustain the effort so long.

Again, there are numerous cases in which no convulsive movements at all can be traced; and where the sudden loss of consciousness, change of colour, and somnolency, can be attributed to the right cause only by a careful analysis of the patient's antecedents and the concomitant symptoms. These are the cases which have been termed by authors, epilepsia syncopalis.

The diagnosis of the affection may here be rendered difficult, and the case may be regarded by the medical man, who has no means of ascertaining the previous history, as a case of apoplexy or of syncope. The absence of stertor has been pointed out by Dr. Prichard as a guide to the affection being epileptic, and not apoplectic. This alone would scarcely, however, be a safe indication, since we meet with cases of apoplexy in which there is no stertor. As it might entail a serious error in the treatment, it would be wise, where doubt existed, to wait for a brief space before applying active measures; as in a case of epilepsy the symptoms would probably, after a short time, assume a different character, which would be sufficiently indicative of the nature of the disease.

The convulsive movements are at times so limited as to remind the beholder of chorea, from which, however, the medical man will readily distinguish them by the fact of their brief duration, as well as by other symptoms. We shall have occasion to see that there is a frequent connexion between genuine chorea and epilepsy, not so much in the same individual as in members of the same family. The younger sister of an epileptic girl who was brought under my notice, was often subject to a peculiar smacking of the lips, which was regarded by the father, himself a physician, as indicating a tendency to epilepsy. Several of my epileptic patients have been subject to spasmodic contractions of one or the other extremity occurring during the intervals of the fits.

Dr. Prichard has devoted a separate chapter to what he denominates partial epilepsy, and details some interesting cases, in which the spasm was confined to a much smaller number of muscles than is generally the case. One instance is that of a man, aged forty-six, who had strong convulsive jerking of the left arm; his arm was violently tossed about for a minute or two, and afterwards felt numbed, and for a short time the patient was incapable of moving it. These symptoms occurred in fits, during which the man never lost his senses, though each attack was followed by vertigo and severe headache. A similar case is detailed by the same author, in which jerking of the right arm gave place to paralysis.

At any time when, under such circumstances, a doubt exists as to the nature of the malady we have to deal with, the cerebral functions should be most carefully inquired into, and the character of their derangement, if any, would determine the question.

Strange symptoms are related by some authors as accompanying the epileptic paroxysm. Boerhaave (quoted by Cooke) met with a Jewish woman who in the fit alternately contracted and elongated her lips; they were thrust out into a sharp beak, and then drawn back with such celerity as to make the beholders giddy.

An erection of the penis (a phenomenon which may be considered under the head of spasmodic action) is stated by Hoffmann to accompany the epileptic seizures of children. It is not improbable that it also occurs in adults, since involuntary micturition, defecation, and seminal emissions, are known to take place.

A spasmodic action of the muscles of the neck is so frequent an occurrence in epilepsy that Dr. Marshall Hall* has based upon it a theory of the disease involving a novel mode of treatment, by tracheotomy. At one stage of the epileptic paroxysm the face becomes flushed, and the muscles of the neck are visibly in a state of excitement and contraction. The flushing deepens in intensity as the spasm increases, and in proportion as the respiratory process is interfered with, it becomes more and more dark, "black" as the patient's friends often say. In one of my patients the limit of the external congestion was so marked as to be described by the observer as being bounded by a straight line drawn transversely across the neck. The internal jugular is most liable to suffer through the spasmodic action of the omohyoid muscle; and it is fair to assume that it would be more exposed to compression than the adjoining carotid, owing to the latter possessing greater resiliency. When the sternohyoid and platysma are together in a state of spasm, the external jugular will also be compressed, and so give rise to the superficial vascular engorgement, which becomes palpably visible in the

---

* On the Neck as a Medical Region, and on Trachelismus, &c. By Marshall Hall, M.D., F.R.S. London, 1849. And On the Threatenings of Apoplexy and Paralysis, Inorganic Epilepsy, &c. By the same. London, 1851.

face. In either case extravasation does at times ensue. An interesting case, in proof of the influence that may be exerted by the two last-named muscles when spasmodically contracted during the paroxysm, has been carefully observed by Dr. Russell Reynolds, and is related by Dr. Marshall Hall in his third essay "On the Neck as a Medical Region."* This eminent physiologist is of opinion that "whatever the violence of the arterial circulation, there is no danger, no tendency to morbid action, as long as there is no impediment to the return of blood along the veins; the idea of a tendency or determination of blood *to* the head is a fiction and a chimera, and the real state of things in the condition which has been so designated is, in fact, its impeded return *from* the head." This impeded return he attributes to the spasmodic action of the muscles of the neck on the veins, "an action evident in a vast many instances, though latent perhaps, and to be inferred from the similarity of its effects in others."

While it would be impossible to deny that spasm of

---

* While using the electro-magnetic machine in the treatment of a case of aphonia, Dr. Reynolds "tried the effect upon the muscles of the neck, and observed that when the wheel was turned slowly, and the superficial muscles were alternately contracted and relaxed, the colour of the face was heightened, and was of a florid hue; but when the wheel was turned rapidly, with a less powerful current, and the muscles were maintained during the rapidly intermitting action, in a state of almost permanent contraction, the face became of a deeper colour, the lips and angles of the mouth livid, the eyes suffused, and some feelings of confusion of thought, headache, and dimness of sight, alternating with flashing of light, were induced. The latter effects remained after the cessation of the current for a few minutes and then disappeared."

the muscles in the neck, the platysma, sternocleido-mastoid, scaleni, and omohyoid more particularly, materially affects the circulation of the blood in the vessels of the part, the literal interpretation and careful observation of all the symptoms of epilepsy will not allow of our regarding this as more than a small part of the phenomena. The manner in which the spasm here acts, upon Dr. Marshall Hall's own theory, is by preventing the return of the venous blood from the brain, and thus causing congestion. But a careful observation of the epileptic paroxysm shows that the spasm of the muscles in question never takes place at the commencement, but only after the lapse of a certain time, which might with propriety be called the first stage of the fit. There are, however, many genuine cases of epilepsy in which there is no evidence of any such spasm occurring. On the other hand the experiments of Sir Astley Cooper, and those of Kussmaul and Tenner, to which there will be occasion to revert more in detail, show that compression of the arteries leading to the head and arterial hæmorrhage, which induce a condition in the brain the very opposite to that resulting from spasm of the muscles of the neck, give rise to epileptic seizures in animals. While there is ample pathological evidence to show that morbid conditions inducing compression of the vessels of the neck generally are frequently associated with epilepsy, cases of an opposite character are not wanting in the history of medicine. The following case, which fell under my personal observation, offers proof of the possibility of cessation of the epileptic fits under circumstances which are ordinarily regarded as liable to produce them. It strongly attracted my attention, as it occurred just about the time that Dr. Marshall Hall put forward his views on trachelismus, and seemed to afford evidence, apart from other considerations, that they did not suffice to establish a *rationale* of epilepsy.

F. T., æt. sixteen, a girl of ordinary stature, well made, without curvature of the spine or thoracic deformity, was in good health previous to her eighth year, except that her eyes were thought defective, from her clumsiness. She was not affected with swelling of any kind; there was no evidence of her having been scrofulous, though her appearance at the time of being first seen by me was heavy, and of a scrofulous character. At eight years of age she was seized with a struggling fit in bed, and became black in the face; she did not scream, and the attack passed off in ten minutes. She continued in her usual health for six months, when she had a second fit at 5 A.M., in bed, lasting six hours; she did not scream, but was perfectly unconscious; every part of the body was agitated, the eyes were turned up, and the neck swollen. After the fit she appeared somewhat deficient in intellect, and did not speak as plainly as before. She then had six fits at intervals of six hours each. The first lasted about six hours: the symptoms were always the same; the attacks always occurred in bed: she had vomiting; the eyes were turned up; there was discoloration of the face. The fits were followed by much flatulence and discomfort. Soon after the fits ceased the neck was observed to swell permanently; there were no further epileptic symptoms; the memory was good, the speech clear, but a connected conversation appeared out of the question. When seen by me she could say her prayers, but had a difficulty in receiving instruction, as she was unable to express what she meant: hearing good, eyesight defective. Since the cessation of the fits there has been constant headache, much augmented during the last three months previous to coming under my care; the patient compares it to earache. Two years ago there was some otorrhœa, the side not known; bowels confined; catamenia regular for a year past. After the second fit the entire left side was paralysed for some time; but at present there is no difference in the strength of the two sides, except a slight strabismus of the left eye. The pupils are sluggish, but contract under the action of light. She never falls, but occasionally stumbles. Measurement of neck above the thyroid thirteen inches and a half, fifteen inches over the thyroid gland at time of consultation; tongue clean; pulse 132, soft. The treatment consisted merely in the application of a seton to the nape of the neck; the exhibition of iodide of potassium, with gentian, and the administration of tepid sponging. The seton rapidly had a very beneficial effect; after wearing it eleven days, the 'earache' had entirely left, and only some frontal headache remained. The last report, about seven weeks after commencement of the treatment, is that there has been no return of the earache; in other respects, no material alteration. As there was much irritation from the seton, this was now removed, and I heard nothing more of the patient.

In contrast with the case just related, I may advert to one given by Dr. Parry,[*] in which the thyroid gland always enlarged to a very great degree before the fits, but returned to its natural state after the fits.

I have hitherto spoken of the spasmodic action of

---

[*] Collections from the Unpublished Medical Writings of the late Caleb Hillier Parry, M.D., F.R.S., vol. i. p. 396. London, 1825.

the muscles in epilepsy as being essentially clonic; but though De Sauvages' definition is true in the main, and is generally adopted, we meet with a variety of epilepsy in which the muscular action shows itself as a tonic or tetanic spasm. It is not, however, to be overlooked that even in the ordinary form of epilepsy the clonic spasm is commonly accompanied by tonic spasm, in the shape of firm closure of the fists, or in what is technically designated as carpopedal contractions. In one of my patients the left arm is stated to be sometimes, as it were, glued to the head during the fit. The general character of the spasm accompanying epilepsy, however, is clonic, and tetanic convulsions must be regarded as the exception. Their occurrence in conjunction with epilepsy has been long noticed. Thus, on referring to Frederick Hoffmann's Consultations,[*] we find, in case 30, that a young woman of twenty-three years presented all the complex symptoms of epilepsy, which are minutely described, but that her body became rigid. Prichard has more particularly directed attention to this peculiarity in certain epileptic patients. He has constituted tetanoid epilepsy as a distinct variety of the disease. In it the paroxysm is essentially such as we meet with in ordinary cases of epilepsy; but, instead of the irregular spasm of the various muscles, the limbs are stretched, and the whole body extended and fixed by a rigid spasm; the eyes are widely open, not reverted, as they are usually, but staring frightfully, and the pupils contracted, and insensible to the stimulus of light.

If, on the one hand, the convulsions show great va-

---

[*] Frederici Hoffmanni Consultationum et Responsorum Medicinalium Centuria prima. Tomus primus. Amstelod. 1734.

riations in character and mode of localization, on the other they manifest extreme differences of degree. We see the spasmodic attack passing through all the changes from a mere twitch of an individual muscle to the most violent excitement of every voluntary muscle of the body. In many cases nothing but a slight spasm is perceptible about the muscles of the neck; in others nothing but a contraction of the fingers or toes—the carpopedal contractions of authors—indicates the nature of the seizure, or the eyes only are peculiarly affected; or, again, a close observer is able to detect nothing of a spasmodic character, and the inference of the case before him being epileptic might be unsafe, unless in the same individual more decided epileptic seizures had been previously manifested, or unless he found that the symptoms were identical with those observed in other instances where the epileptic nature was undoubted.

It is common to distinguish from epilepsy a form of disease termed eclampsia, in which the main symptom is a temporary loss of consciousness, not of a syncopal character, and not attended with any spasmodic action of the voluntary muscles. I confess that the gradations by which the features of well-marked epilepsy pass into those of eclampsia, appear to me so imperceptible, that I doubt the propriety of distinguishing the two diseases nosologically; I should be disposed to regard eclampsia as a variety of the same type of disease, but characterized by the absence of spasm of the voluntary muscles.*

---

* On the subject of the transition of different forms of spasm, and the impossibility of always drawing an exact line of demarcation, I would refer the reader to Dr. Copland's admirable remarks: Dictionary of Medicine, vol. i. p. 218, seqq.

CHAPTER III.

Analysis of individual symptoms continued — Biting the tongue — The pulse in the paroxysm — Frequency of the fits — Periodicity — Lunar influences — Dr. Mead's case — Nocturnal and diurnal influences — Dr. Boyd's observations — Seasonal influences — Dr. Moreau's observations — Headache — Cases of epileptic headache — Delirium and hallucinations.

BITING the tongue and the inside of the mouth or lips is a symptom of some importance in the diagnosis of epilepsy, because it indicates that the unconsciousness must have been complete and the spasm violent. The absence of the symptom by no means proves a seizure not to have been epileptic; for it may be regarded as a mere accident, like any other injury in the epileptic, induced by the coincident convulsion and unconsciousness, as a fall into the fire or against a sharp corner. In cases of doubt the fact of the tongue being bitten would be strong evidence in favour of the view that the case under consideration was one of epilepsy. In my first analysis of fifty-two cases of epilepsy I found it to have occurred seventeen times, or at the rate of 32·7 per cent. In my second series of fifty-two cases it presented a greater frequency; twenty-eight patients, or 53·8 per cent., having bitten their tongue to a greater or less extent. If we take the two series together, we find that of 104 patients forty-five bit their tongue, or 43·27 per cent. In some patients the tongue is bitten at every fit; in others it is perhaps injured occasionally; in others it may be bitten once, and not again. Thus in the second series of cases, examined with reference to this point, I find that there were three who only bit their tongue in one fit, their other attacks not being marked by this symptom; whereas, in all the others, it occurred frequently. We may not always expect to find a trace of the injury, for the extent of the lesion varies much; and even when a considerable solution of continuity takes place, it is remarkable how soon and completely the repair is effected. As in the same patient the convulsions are often noted to prevail on one side of the body, so also do we find that the injury done to the tongue is commonly limited to the right or the left side.

The views of Schröder van der Kolk* regarding the limitation of the irritant cause which induces the epileptic paroxysm in its action upon certain portions of the

---

* See his work On the Minute Structure and Functions of the Spinal Cord and Medulla Oblongata, translated by W. D. Moore, M.B., London, 1859, p. 238, seqq. Dr. Schröder van der Kolk has found that in patients who had habitually bitten their tongue the capillary vessels in the course of the hypoglossal nucleus in the medulla oblongata, presented a preponderance in their diameter over those in the track of the vagus, and he connects the two as cause and effect: "For," he adds, "as this phenomenon, although very frequent in epilepsy, does not occur in all cases, it appeared to me that if, during the fit, the vessels which are so much wider, conducted more blood to the nucleus of the hypoglossus and the corpus olivare, which is in such close connexion with it,—it was not improbable that these parts would be more irritated, which might cause convulsive movements of the tongue, and the protrusion of this organ between the teeth, and thus occasion it to be constantly bitten during the attacks of epilepsy."

medulla oblongata, according to the presence or absence of the symptom in question, receive a general confirmation from my experience. At the same time it is manifest that even in the same individual the root of the hypoglossus in the fourth ventricle may at one time be irritated, and not at another.

A few words will comprise all that may be said about the pulse during the paroxysm. There is no feature in the pulse which is in any way pathognomonic. The general circulation suffers secondarily from the attacks, and shows greater or less excitement, proportionate to their violence. The pulse is accelerated, and its force varies with the general vigour of the patient.

During the convulsions the pulse is at first excited, and more or less increased in force and frequency; when from continued and violent spasm exhaustion sets in, the pulse may become imperceptible, and lead to the apprehension—as I have before remarked—of death having taken place. During the recovery from an attack the pulse continues for a time to exhibit morbid excitement, which is gradually allayed until it recovers its usual habit.

During the intervals of the fits its force and frequency present the usual variations which are found to be associated with greater or less physical strength. The prevailing character of the pulse in the majority of cases that have fallen under my observation, is that which is met with in subjects who want blood and tone. It is generally much accelerated, feeble, and soft. In cases exhibiting more of the sanguineous and florid type, the pulse may present no variation from the normal character. But this is rarely the case, and this frequent quickness of pulse is probably the reason, as it certainly is an occasional useful indication, for the employment of digitalis. The amount of digitalis which some epileptic patients can take before a perceptible influence is exerted upon the pulse, is at times surprising. A normal pulse will occur more frequently when the attacks have only commenced recently, or occur at long intervals; but even here there are exceptions, as in the case of a gentleman who was sent to me by Mr. Davis, of Heytesbury, who although for fifteen years subject to epilepsy occurring once in three or four weeks, had a pulse only 68 and full, when I saw him. The more frequent the attacks are, the more perceptible will be the derangement of the circulation; but in no case does it appear that even prolonged cases of epilepsy exercise any definite influence upon the central organ of the circulation. Nor can any causal relation be traced, such as exists between chorea and morbid conditions of the heart. In short, the manifestations of any influences of epilepsy upon the circulation appear to be confined to such effects as may be found in any circumstances which at once enfeeble the individual while they excite the circulation. An exception from the general rule that an acceleration of the pulse is produced by epilepsy, if any change is perceptible, is detailed by Dr. Burnett,* who quotes two similar instances from Morgagni, as the only ones on record. In Dr. Burnett's case, an officer, aged forty-six, became epileptic, and the pulse was reduced from the normal standard to 20, and at times sunk as low as 14 in the minute during the fits. This slowness became persistent, and is attributed by the author to

---

* Case of Epilepsy attended with remarkable Slowness of Pulse. By William Burnett, M.D. (Med.-Chir. Trans., vol. xiii. p. 20.)

derangement of the chylopoietic viscera, which appears also to have been the cause in the cases related by Morgagni.*

The frequency of the fits and the question of their periodicity will next command our attention.

The complete paroxysm, of which alone I have hitherto been speaking, occurs at very varying intervals; sometimes a single fit may occur never to return; sometimes the second fit may not show itself, as in one of my own cases, for seventeen years after, and then a rapid succession of attacks may take place. The more ordinary case is, that at first the fits present themselves at intervals of a few months, and gradually increase in frequency, until we find them occurring day by day, and even repeatedly in the course of twenty-four hours. The general tendency of the disease, if left to nature, is certainly to go on from bad to worse, and not to terminate in a spontaneous cure. In some cases, a constant and rapid succession of epileptic seizures will affect the patient, with scarce a moment's perfectly free interval: it is in such circumstances that death is to be feared, and does take place, as I have myself witnessed. But neither in my own observations, nor in the histories of the disease preserved by other authors, has there appeared to be any uniformity in the mode in which the paroxysms took place. An approach to regular periodicity is sometimes observed in the female, as the disease there, at times, bears a palpable relation to the catamenia; but this is by no means uniform nor sufficiently marked to lay down a general law. Moreover, in this case the periodicity can scarcely be said to be a feature of the epilepsy; but the semblance of a

---

* Book i. letter ix. art. 7; and letter lxiv. art. 5.

periodical return is due to the spasmodic paroxysm depending upon another condition, which in its nature is of a periodical character.

From the earliest ages a different doctrine has been advocated. Lunar influences are still upheld by some physicians of repute, and the belief is generally in vogue among the public, though it has been combated by Hippocrates and most of his successors. It would be difficult to account for the persistency of a fallacy for which there is but a very slender support, were it not that so few persons prefer to inquire for themselves rather than take a dictum upon some acknowledged authority.

The periodicity to which I allude as having been supposed to prevail in epilepsy, was attributed to the moon; hence epileptics were termed moony or moonstruck (σεληνιαζόμενοι). Demoniac possession appears a much more intelligible doctrine to any one who has once witnessed the terrific contortions of some patients; but for the influence of the moon we can find neither a parallel in human pathology, nor a proof of its reality in the disease before us, satisfactory to our own minds.

A very interesting case is related by Mead, in which a periodicity was stated to occur, which countenances the belief in the influence of *Selene*; but its very rarity justifies our seeking for some other explanation, while I am able to bring forward authorities which, apart from my own experience, appear to be sufficiently conclusive on the point. Dr. Mead,* who though a distinguished physician in his day, was an astrologer, and, therefore,

---

* The Medical Works of Richard Mead, M.D. London, 1762. One of Dr. Mead's works is entitled, "De Imperio Solis ac Lunæ in Corpore Humano."

naturally a firm believer in the moon's power in causing epilepsy; with Galen, he was of opinion that the moon governs the periods of epileptic cases, and he states that he has often predicted the times of the "fits with *tolerable* certainty."

The case in question is always quoted as one of genuine epilepsy. I cannot, however, avoid remarking that it is a very unusual feature in this disease that the paroxysms should terminate with a loud cry; still as the patient was too young either to simulate or to labour under hysteria, the coincidence of the affection with the tides remains a curiosity in medical literature, even allowing for the author's prejudices.

The following is the account of the case referred to:—" But no greater consent in such cases was, perhaps, ever observed, than what I saw, many years since, in a child about five years old, in which the convulsions were so strong and frequent that life was almost despaired of, and by evacuations and other medicines was with difficulty saved. The girl, who was of a lusty, full habit of body, continued well for a few days, but was, at full moon, again seized with a most violent fit; after which the disease kept its periods constant and regular with the tides. She lay always speechless during the whole time of flood, and recovered upon the ebb. The father, who lived by the Thames' side, and did business upon the river, observed these returns to be so punctual, that not only coming home he knew how the child was before he saw it, but in the night has risen to his employ, being warned by her cries, when coming out of the fit, of the turning of the water. This continued fourteen days—that is, to the next change of the moon; and then a dry scab on the crown of the head (the effect of an epispastic plaster, with which I had covered the whole occiput in the beginning of the illness) broke, and from the sore, though there had been no sensible discharge this way for above a fortnight, ran a considerable quantity of limpid serum; upon which, the fits returning no more, I took great care to promote this new evacuation by proper applications, with desired success for some time; and when it ceased, besides three to four purges with mercurius dulcis, &c., directed to be taken about the new and full moon, I ordered an issue in the neck, which, being thought troublesome, was made in the arm. The patient, however, grew up to woman's estate without ever after feeling any attacks of those frightful symptoms."

I will not weary the reader by a quotation of the authors who have written on both sides of the question. The doctrine of the moon's influence is, however, by no means extinct.* The question is one of a very vague character; and the answer is manifestly not as easily attainable as, for instance, to an inquiry relative to the causation of the tides, otherwise we should not still find it necessary to enter upon the inquiry. Some medical men continue to hold to the lunar influence, and among the lay public we often meet with evidence that the belief still prevails. Thus the mother of Mary Ann A. (one of my patients) stated of her daughter, that for three years she had been subject to fits "at the full and change of the moon." I have received similar indefinite statements from a few other patients, but have never succeeded in eliciting any satisfactory evi-

---

* Romberg (l. c. vol. ii. p. 205) does not state his own experience on the subject, but says that, "although here and there doubts have been raised against this view, the accurate observations of others have established its correctness."

dence of the alleged periodicity. Nothing but a most minute and extensive analysis of a large number of epileptic seizures could satisfactorily determine, whether this disease bears a relation to the phases of the moon different from what is observed in other morbid conditions.

Dr. Moreau* has, in his prize essay on Epilepsy, set the question at rest. The author, whose position at the Bicêtre gave him the best opportunities for instituting a rigorous inquiry and insuring correct reports, analyses 42,637 attacks occurring successively in one hundred and eight male patients in the course of five years. For the details of the analysis, I refer the reader to the Memoirs of the Academy of Medicine; it will suffice for the present purpose to give the general results. The 42,637 were thus distributed: Between the phases of the moon the number of epileptic seizures was 26,313; the number occurring during the changes themselves was 16,324. The difference in favour of the former was therefore 9984; or the relative frequency of the occurrence of the fits at the changes and in the intervals was as 16 to 26. Dr. Moreau concludes, I think justly, that the changes of the moon exert no influence upon the epileptic seizures, as they are more frequent during the intermediate periods.

The tendency of the human mind to adopt foregone conclusions is often shown in the manner in which persons, favouring the notion of a lunar influence in epilepsy, interpret the very frequent exceptions from the

---

† De l'Etiologie de l'Epilepsie. Par le Docteur J. Moreau (de Tours), Médecin de l'Hospice de Bicêtre. Mémoire couronné.—Mémoires de l'Académie de Médecine, tom. xviii. Paris, 1854.

PERIODICITY OF EPILEPSY.

coincidence between the fits and the lunar changes in those patients in whom they watch the influence of the latter with solicitude. Thus such persons will view a fit occurring two or three days before or after the full or the new moon as coming within the range of such influence; but it must be manifest that, if we seek for a definite relation between the epileptic seizure and the moon, according to the analogy of the tides, we should repudiate all conclusions that are not based upon similar clear and irrefragable evidence. I have myself no bias beyond that induced by the facts that are available. I certainly do not seek for what is commonly regarded as the marvellous; but this, not because I am unwilling to receive what I cannot fully comprehend, but because, being in daily life surrounded by wonders, it is unphilosophical to multiply them where the ordinary processes of reasoning based on statistics show them not to exist.

A more definite periodical type is observed in a large number of patients in the cycle of the twenty-four hours, the fits very commonly showing a preference either for day or night in the same individual.* The circumstance has nothing in common with the type of intermittent diseases, but depends upon the physiological effect which sleep, the recumbent position, the emptiness or repletion of the stomach, the state of the sexual functions, &c., produce upon the brain. There is a peculiar proclivity in some persons to nocturnal attacks; at least there is a marked difference in the frequency with which some persons are attacked at

---

* A patient's "night" begins and ends according to the period he spends in bed; his day is the time not occupied in bed.

night. They are seized by the convulsions while asleep, or they wake up first and are then attacked. In some the fits occur exclusively, or more often early in the morning, shortly after rising, than at other times. As the nature of the changes that take place in the cerebral circulation during sleep are themselves hypothetical, their influence in the production of epilepsy must necessarily be so also. It does not, however, appear illogical to assume that during sleep there is an increased afflux of blood to the head, both from the prone position, which is well known to favour its production, as from the analogy with the effects produced by narcotic poisons. Such, at least, is the opinion of most authors who have touched upon the subject. In a practical point of view it becomes important to bear this influence of sleep in the production of epilepsy in mind, since it serves as an indication in the treatment. In proof of the increased local afflux of blood, I would cite the benefit which is often obtained by the application to the head of cold lotions on retiring to rest, and the prosecution of similar measures calculated to maintain a due balance in the circulation. By this remark I do not wish to shelve the question of the causation of epilepsy; I would rather anticipate a misconception by at once stating that during sleep, as well as at other times, gastric derangement, and other influences, often act as exciting causes of the paroxysm.

In general hospital or private practice it is difficult to obtain statistics that are sufficiently precise to determine the question of nocturnal and diurnal influence.

In drawing upon the valuable Reports of the Somerset County Pauper Lunatic Asylum, prepared by Dr. Boyd, for information upon this point, I would take an opportunity of expressing a regret that an important and positive gain which might be secured to medical science is wasted by scientific matter being inserted in reports addressed to lay governors of hospitals and charitable institutions, who, with rare exceptions, are unable to appreciate it. It would be no mean undertaking if individuals, or societies like the Sydenham Society, would seek to rescue from unmerited oblivion some of the valuable facts thus attainable.

But to return to Dr. Boyd.* Forty epileptic subjects who were under his observation in 1852, and whose seizures he has analyzed, had 3202 fits, 1962 of which occurred by day and 1240 by night. In 1853 there were forty-nine cases of the disease, in twenty-four males and twenty-five females, who had 3998 fits, 2407 occurring by day and 1591 by night.

In the annexed table (No. 1, p. 50) is exhibited the relative frequency of fits by day and night for every month in the year 1852.

The average of attacks by day and night throughout the year 1853 is given in the second table.

From the remarks which follow the tables, it will appear that, according to my experience, there is a palpable distinction, at least during the earlier stages of epilepsy, between the patients who have fits only by day, and those who experience them only by night. I believe that this distinction is rarely quite obliterated, even where the disease is of long standing. This must be borne in mind in estimating the value of Dr. Boyd's statistics.

It would appear that in Dr. Boyd's cases the seizures affected the same individuals by day and by

---

* See the Reports of the Somerset Lunatic Asylum for 1853 and 1852.

No. 1.

	Jan.		Feb.		March.		April.		May.		June.		July.		August.		Sept.		Oct.		Nov.		Dec.		Total.
	M.	F.	M.	F.	M.	F.	M.	F.	M.	F.	M.	F.	M.	F.	M.	F.	M.	F.	M.	F.	M.	F.	M.	F.	
1852.																									
Day	43	106	106	161	52	171	42	128	68	86	51	94	57	66	65	72	48	124	35	53	73	72	35	154	1982
Night	24	108	35	179	49	112	23	35	42	31	41	31	24	32	27	35	57	72	50	30	54	38	28	85	1240
1853.																									
Day	54	73	47	69	88	70	53	115	61	91	58	106	57	171	118	157	104	135	116	107	134	231	97	93	2407
Night	28	39	34	25	87	57	120	70	111	41	84	65	67	62	79	107	50	53	43	65	75	142	41	51	1591

No. 2.

	Day.	Night.
Males	52	43
Females	76-7	42

night. As they were necessarily patients labouring under the disease in an advanced and aggravated form, in most instances probably complicated with insanity, they would occupy a different position from those who fell under my observation, and who for the most part had been liable for a comparatively short period only. I have analyzed fifty-two of my cases with reference to the diurnal variations, and find that, excepting eight, all afford information on the point.

The largest number occurred by day—viz., nineteen; and of these it is stated that thirteen occurred exclusively by day, or at particular times of the day; only eight occurred indiscriminately by day and night; while seventeen patients averred that their seizures occurred chiefly or exclusively by night; six of these stated positively that they were only liable to night attacks. But more than this, some of the patients always suffered in the morning, others only in the evening, and though I have not succeeded in all cases in finding out an adequate pathological explanation of these peculiarities, it is a subject of great importance and full of interest as bearing upon the causation of the disease. More frequently the time appeared to bear some relation to the patient's meals than anything else, and exhaustion from taking them irregularly or at too long intervals seemed to play an important part in the production of the attacks. Again, when the fits occur exclusively at night, or early in the morning, we must pay especial attention to the sexual organs, as in both sexes sexual irritation is a common cause of epilepsy.* Nor may we

---

* As this point has not been much attended to by writers, I annex seriatim the reports of the cases as to the diurnal period of the occurrence of the attacks:—1. Not stated. 2. Chiefly

52. EPILEPSY AND EPILEPTIFORM SEIZURES.

overlook the influence of defective oxygenation as a possible cause of nocturnal attacks.

It has been stated by some authors, and it is a popular belief, that certain seasons exercise an influence in the production of epilepsy, and that it is more frequent in spring than at other times. On this question no one can afford more definite information than M. Moreau. His analysis of the entire number of fits occurring in 108 epileptics in the course of five years yields the following per-centage for the four seasons :—

Summer	25·2
Spring	26·8
Autumn	23·4
Winter	24·3

The actual number of fits occurring in each month is stated by Dr. Moreau to have been :—

---

morning. 3. Nightly. 4. Chiefly at night. 5. Ditto. 6. Only by day. 7. By night and by day. 8. Generally at night. 9, 10, 11. Not stated. 12. Chiefly at night. 13. Never at night. 14. For three years the fits occurred only by day, subsequently by day and at night. 15. Chiefly at night. 16 and 17. Not stated. 18. At night. 19. Not stated. 20. Mostly by day. 21. Fits occur in the evening, but never at night. 22. Night and day. 23. Chiefly at night. 24. Day and night. 25. Night. 26. By day. 27. Only by night. 28. By day. 29. By night. 30. By night. 31. Chiefly by night. 32. By day. 33. Not mentioned. 34. By night only. 35. Never at night, irregularly by day. 36. Generally by night. 37. Always before twelve o'clock in the day. 38. Always about breakfast time. 39. Generally by day and before dinner. 40. By day. 41. Chiefly at between five and six in the afternoon. 42. At night. 43. Day and night. 44. Day and night. 45. Almost always early in the morning, never at night. 46. By day. 47. By night and day. 48. Always before midday, generally at six in the morning. 49. By night and day equally. 50. Chiefly at night. 51. Chiefly in the morning on rising. 52. By day.

INFLUENCE OF SEASONS—HEADACHE.

January	3944
February	3709
March	3749
April	3732
May	3972
June	4025
July	3657
August	3081
September	3131
October	3472
November	3426
December	2739
Total.	42,637

These numbers give some colour to the popular notion derived from Hippocrates that spring is the favoured season; the difference is, however, not as much as 1½ per cent., and cannot, therefore, deserve very serious attention in reference to the etiology or treatment of epilepsy.

Headache is one of the symptoms that bears so close a relation to the epileptic paroxysm, that it deserves a separate consideration; the more so as the frequency of its occurrence may become an important argument in the consideration of the proximate cause of the disease. It has not hitherto met with quite the consideration which it deserves.

In my own cases I have met with headache in fifty-six out of 104 cases, or in the ratio of 53·8 per cent. This is a lower ratio than what was deduced from the analysis given in the first edition, because in the second series of cases recently presented to the Medico-Chirurgical Society the proportion was much smaller than that yielded by the first. This is a proof, if any were wanting, of the necessity of not placing too absolute a reliance upon the statistics of any individual observer, as they are always liable to correction by a larger field of experience.

The pathological import of the headache varies much according to the period at which we meet with it; because it may be connected with the fit etiologically, or it may be a consequence of the attack, or again it may be a mere casual coincidence.

If we examine the above fifty-six cases in which headache occurred, we find that it was constant or frequent in twenty-eight cases; it occurred before the fits only in four cases; after the fits only in twenty-two cases; and it occurred both immediately before and after the seizures only in three cases. The significance of the headache occurring immediately after the fit only, is necessarily distinct from that of the cephalalgia preceding the attack, or habitually affecting the individual. The violence of the convulsions, the spasmodic affection of the muscles of the neck, the general excitement of the circulation, conspire to exhaust the nervous system, and to induce the conditions in which, under other circumstances, we very commonly meet with headache. Hence it would be strange if the direct result of an epileptic fit were not often to bring on this symptom. It cannot by itself be taken as an indication of the nature of the disease. The headache that habitually affects the epileptic, or that precedes the seizure, occupies a different position, and forms a material item in our arguments regarding its proximate cause. In all these cases, the concomitant state of the circulation, of the chylopoietic viscera, the general state of nutrition, and of other cerebral symptoms, must be taken into account, to enable us to judge of the exact value to be put upon the symptom, and the extent to which it merits consideration in our therapeutic proceedings. To these points we shall revert in a future chapter.

In the same way as epilepsy is very commonly associated with cephalalgia, so also do we frequently meet with individuals who are not epileptic, subject to headache; but in whom the concomitant symptoms of giddiness, and temporary and partial loss of power, remind the physician of their possible relation to epilepsy, or of the approach of the latter disease. Where we are unable to detect any visceral derangement to account for such symptoms in young persons, we must be on our guard. This, of course, will be the more necessary if at any time there have been an epileptic seizure. In classifying different forms of headache I have thought the introduction of the term cephalalgia epileptiformis necessary to express a pathological fact. On this point I have, on a former occasion,* made some observations which I may be permitted to repeat here:—"Among the causes of cephalalgia residing, as it were, in the cranial contents, or affecting them without the intervention of other organs, the foremost are mental and intellectual excitement carried beyond the normal limits of healthy stimulation; they may act by a shock, as in a child that is frightened or suddenly carried into a brilliant light, or made the gazing-stock of pseudo-admiring friends at an hour when it ought to be in the land of dreams; they may act by the continued strain upon the

---

* On Chronic and Periodical Headache; in the *Medical Times and Gazette*, August 12th, 19th, and 26th, 1854; see also Tables of Analyses of One Hundred Cases of Headache, *Association Medical Journal*, Nov. 9th and Nov. 16th, 1855, by E. H. Sieveking, M.D.

reproductive energies of the earnest student diving into the secrets of nature, that are revealed to none but such as are prepared to sacrifice themselves on her altars; it may, after days and nights of unremitting toil, prostrate the poor milliner who, with a weary heart and aching eyes, uses her very life-strings to earn bread for her invalid parents. A direct stimulus is anything that, through the organs of smell, of hearing, or of sight, gains admission to the perceptive faculties. The excitement may produce but a temporary result commensurate with the powers of the individual; or it may find a favourable nidus; and the deranged action once set up, may be propagated until broken by some greater countervailing power. Many forms of epilepsy are nothing but the spasmodic expression of this derangement; and for private use I constantly employ the term *cephalalgia epileptica*, as indicative at once of what I regard as a cause and a tendency, occurring possibly in a subject in whom the epileptic paroxysm has been manifested merely by slight vertiginous attacks, by a single attack in former times, or by some spasmodic action that alone would not be regarded as of an epileptiform character. The pain in this form of headache may affect any part of the head, but it is frequently limited to a spot at the vertex; and where that is the case, I have found marked benefit arise from making the attack directly upon the apparent seat of injury." In an analysis of one hundred cases of headache, taken successively as they presented themselves in my memoranda, and published in *The Association Medical Journal*, two cases of epileptiform headache are given; and although the number is scarcely sufficient to serve as a basis for calculations, I think that two per cent. does, judging by my reminiscences, approximatively indicate the frequency of this form of the complaint.

The following, which is not one of the two cases adverted to, is an instance of what may be called cephalalgia epileptiformis :—

E. G., a widow, aged thirty-seven, of robust appearance and florid complexion, has always enjoyed good health, except that she has been subject to headache, supposed to be connected with the fact that her mother's head was "split open" when she quickened with patient. A year before consultation, E. G. felt a sudden numbness in her right leg, ascending to the trunk, right arm, and face, with a film over her eyes, and leaving a violent headache, lasting the whole day. The numbness passed off after two hours. These attacks returned about once a month; they took away her senses, but not to such an extent as to prevent her being conscious of what was passing around; articulation became impaired, and the patient complained of being very nervous.

I may be permitted to draw upon my note-book for another instance of a similar kind, as illustrating the occurrence of the affection in the child. It is not one of those included in the table.

M. I., aged four, the daughter of a painter, had inflammation of the brain a year ago. Was unconscious for a week. The illness lasted a fortnight; but she was neither blistered, bled, nor leeched. Two years ago, without known cause, she had fits, but they have not since returned. For a year the child has been subject to a pain in her temples and occasional vertigo; she turns pale, and sits down, and says she is sick, but does not vomit. The pain is paroxysmal. She loses the use of her limbs at times, at night, when taken up for certain wants. There were no carpopedal contractions. There was evident derangement of the primæ viæ, which was regulated by medicines and diet. The prolonged exhibition of steel wine effected a complete cure of the headache.

It seems a just inference that the causes which induce the headache that follows the epileptic fit are also indictable for the somnolency that, with but few exceptions, ensues after the paroxysm has passed off. In some cases the patient at once returns to his ordinary state of existence, but in the majority of instances he falls into a profound sleep, which lasts a shorter or longer period. Sometimes the drowsiness is considerable, but the patient is able to rouse himself if necessary; again, there may be no perceptible interval between the fit and the subsequent sleep, the patient passing insensibly from the abnormal condition into that of normal sleep. There is commonly a feeling of exhaustion after a fit which is not necessarily proportionate to the violence of the paroxysms; the mere fatigue resulting from the spasmodic action might otherwise be sufficient to account for the sleep; but this relation is not sufficiently definite, though undoubtedly the amount of muscular action has some influence upon the subsequent sleep. On the one hand, we at times meet with violent epileptic convulsions which are not followed by sleep; thus in a girl aged ten, who for three years had been subject to violent epileptic fits, and whose mind was gradually giving way under their influence, no sleep ensued after them, and her sleep was generally better at the times when she was free from the seizures; on the other hand, we meet with cases in which the paroxysm is very slight, and the stupor and drowsiness strongly marked.

Instead of the sleepiness, we sometimes observe a state approaching to or constituting actual delirium; or the same patient at one time sleeps a long time after the fits, and then a period comes when he or she is delirious for several hours after the paroxysm has gone off. In the same way hallucinations are found at times in conjunction with epilepsy, though these occurrences are to be reckoned exceptional.

Since the publication of the first edition of this work, the profession has lost one of its brightest ornaments, Professor Alison, of Edinburgh, in whose case hallucinations and delirium were associated with epilepsy. In the account* given by Dr. Patrick Newbigging of the illness and death of this eminent physician and teacher, we find that his first seizure in 1846 was followed by delirium of a violent, and, for such a man, highly demonstrative character. The fits recurred at intervals of six weeks till 1850; they then became more frequent and severe, some being followed by headache and nervous excitement. In 1854, after a severe seizure in the month of October, Dr. Alison suddenly became delirious; the delirium passed off on the following day, and the patient recovered his usual state of health. During the next five years the frequency of the fits fluctuated a good deal. " The nature of the terrible attacks," to employ the author's words, " to which Dr. Alison had become so great a martyr, gradually underwent certain changes. They became less numerous, but the fits were greatly increased in severity, and during this year the condition of my patient after the fits was very formidable and most painful to witness—the delirium assuming more and more the maniacal

---

* Edinburgh Medical Journal, Jan. 1860.

character, and latterly the difficulty of restraint, even with the assistance of two or three male attendants, was very considerable." Towards the close there were frequent changes from comparative quietness to complete maniacal excitement until within a few days of dissolution, when Dr. Alison's state was thought to resemble that of a person sinking under typhoid fever. "There were brief intervals towards the close," says Dr. Newbigging, "when our patient was sensible, was able to express his thanks to those around him in his usual calm manner, and when he stated his belief that his end was approaching. But these peaceful moments were soon followed by excitement characterized by violent spasmodic action and screaming, accompanied by a peculiar rotatory movement of the hands, or by a condition very frequent in the earlier attacks, at a certain stage of the seizure, when he seemed lost in the contemplation of some blessed vision, during which he expressed his belief that he heard the praises of the heavenly hosts, and that among the number he distinctly recognised the voices of dear departed friends."

CHAPTER IV.

The phenomena observed during the free intervals — State of chylopoietic viscera — Pupil — Vertigo — Petit mal — The sequelæ of epilepsy — Loss of memory — Fatuity — Paralysis — Moreau's statistics — Impaired circulation — Derangement of intellect — Fatality of epilepsy — Statistics of Registrar-General — Author's calculations.

IN my opening remarks I have observed that the violence and peculiar character of the epileptic paroxysm very commonly cause symptoms to be overlooked which, occurring in the intervals, and not palpably bearing any immediate relation to the fit, are therefore regarded as offering no aid in the interpretation of the disease. The tendency, and I think the fault, of medical science at the present day is to divide and refine too much; one is apt to forget now-a-days that the human body, though a complex of various organs, is essentially one and indivisible, and that no part can be well understood except in its relation to the whole. The most egregious errors are but too frequently committed when the practitioner seeks to treat the disease of a part without remembering its interdependence with other parts; and the individual patient, no less than the science of medicine itself, suffers from the prevailing specialism. In the same way as the disease of a limb or organ necessarily reacts upon, and is reciprocally influenced by, the condition of the entire body; so, too, may we not separate the different phases of man's existence from one another, setting them up like signposts in a wilderness, without an indication of where they point the traveller's footsteps to? Rather must we ever seek to read the entire man, not forgetting how interdependent all his vital phenomena are; and especially how closely interwoven with all his physical acts are the thoughts and feelings, sufferings and rejoicings that emanate from or manifest his spiritual existence. These remarks will receive frequent illustration when we arrive at the consideration of the influences to which epilepsy must often be attributed; at present they are submitted to the reader, in order to enforce, in the appreciation of the epileptic paroxysm, the necessity of not separating it from the intervening periods, or so-called free intervals. In order to understand the nature of epilepsy, a more careful study of the general condition of each patient, and especially of the symptoms that may show themselves in the free intervals, is necessary. Epilepsy is a disease of the whole man, and not of any one organ or system of organs alone.*

At times, as we have before remarked, a single paroxysm may occur, never to return. This, however, is rather the exception than the rule. Still the fact must be regarded, to whatever immediate cause the individual attack was attributable, as indi-

---

* In a very friendly and flattering review of the first edition of this book, contained in the *Asylum Journal* for April, 1858, the writer objects to this passage on the ground that epilepsy is essentially a disease of the nervous system. I allow the sentence to remain in its original terseness, because, taken in the context, it upholds a view which I still adhere to. I do not, of course, deny that the nervous system, and particular portions of it, are mainly at fault in this disease. In the chapter on the theory of epilepsy and elsewhere, much evidence in corroboration of

cative of a peculiar predisposition in the person. The more common circumstance is that, one paroxysm having taken place, after the lapse of one, two, or three months, it is repeated; that then a period of the same length transpires until the occurrence of the third attack, or that the period is abridged, and that again a comparatively long free interval ensues, followed by a fourth attack, and so on: several intervals may present the same duration, and then a shorter free period may intervene; but the general tendency is to a diminution of the interval, and a corresponding increase in the frequency of the attacks. The more frequent the seizures, the more serious the aspect of the disease. Like ague, the shorter the intermissions, the firmer is the hold which the morbid condition has upon the system; hence we cannot but regard it as beneficial and promising if, as a result of therapeutic proceedings, we find the space that separates one fit from another gradually widening. The more frequent the fits, the more marked will be the symptoms of disease which may be traced in the interval; and yet a careful observer will rarely fail to discover a certain peculiar deviation from health in epileptic patients, even if the intervals are protracted. There will be the characteristics of a nervous diathesis; an excitable, frequently

---

this point will be adduced. But I wish to convey that we cannot sufficiently estimate the relations of the nervous phenomena, and determine the most scientific and satisfactory mode of treatment, unless we have regard to "the whole man," and avail ourselves, on the one hand, of all the aid physiological pathology affords us for the investigation of disease, and, on the other, of all the avenues by which hygienic and medicinal treatment may influence its arrest or alleviation.

irritable, manner; a restless eye, a quick but feeble pulse; there is more or less difficulty in collecting the thoughts and connecting the different links of mental association, while at the same time one or other of the organic functions presents a palpable deviation from health. The organs that are more particularly under the domain of the sympathetic ordinarily show that they are deficient in vigour, that they want that stimulus which the vascular and nervous systems supply when the individual enjoys robust health. Hence a common symptom is a torpid state of the intestinal tract, as shown in flatulent dyspepsia, eructations, intestinal flatulency, and constipation.

An associated symptom is an enlarged and sluggish state of the pupil, such as is commonly met with in persons suffering from the presence of intestinal worms, from a morbid condition of the generative organs, or from a torpid condition and enlargement of the mesenteric glands. I have, however, though rarely, seen a contracted state of the pupil. In one case—which is so remarkable in other respects that I shall quote it at length in a future chapter—the contraction of the pupils came on so regularly before the fits, that the mother, a very observant lady, came to rely upon the symptom as a sign of an approaching paroxysm.

Occasional vertigo; irregular, frequent, or constant headache, with or without vertigo, and not traceable to any definite exciting cause; anomalous sensations in different parts of the body; slight partial spasmodic seizures, more particularly a distressing sense of suffocation or choking, belong to the symptoms commonly met with in the free intervals. The suffocating sensation, last alluded to, is often identical with the description given by patients of the globus hystericus, and no doubt depends upon the same nervous condition; I have, however, observed in decidedly epileptic cases that the sensation is more distinctly described as that of constriction round the neck, than as the "ball rising up the throat" of hysterical females. The choking sensation is observed in males as well as in females. Dr. Parry,* who forcibly dwells on the close relation of various nervous affections, particularly adverts to this point, and expresses the opinion that globus hystericus, epilepsy, and mania, are but different gradations of the same fundamental affection. I have already quoted instances proving that the paroxysm itself varies in the completeness of its symptoms; numerous instances are given by most authors who have written on the subject, showing the infinite gradations that are met with, from the complete and violent epileptic fit, to the merest twitching of a single muscle. Esquirol† relates that some epileptics only shake the head, the arm, the legs; others only close the hand, run or turn round. Dr. Esparron recognised an attack of epilepsy by a simple convulsive movement of the lips; Pechlin drew the same conclusion from convulsions of the eyes and thorax. I have at times been at a loss to give a distinct name to certain symptoms manifested in my patients, until the outbreak of an epileptic paroxysm declared their nature and showed their relationship. The following may be quoted from my case-book as an instance:—

E. H., a married woman, æt. fifty-four, who had

---

* Collections from the Unpublished Medical Writings of the late Caleb Hillier Parry, M.D., vol. i. p. 392, seqq. London, 1825.

† Des Maladies Mentales, &c., tome i. Paris, 1838.

previously been in perfect health, was seized once a week, or at longer intervals, with occasional loss of speech and tremors, lasting for half an hour or more; these attacks were accompanied by pain in the head and shoulders; she very rarely lost her consciousness. These seizures were followed by great weakness and prostration. After the lapse of a year they became severer, and they now assumed the complete character of epilepsy. There now was always entire unconsciousness, foaming at the mouth, gnashing of the teeth, and convulsive movements.

Many patients, after the outbreak of the typical paroxysm has removed all doubts as to the nature of the disease they are labouring under, suffer in the way that the subject of the last-quoted case did before the appearance of the epileptic fit. The French distinguish between a *grand mal* and a *petit mal;* the former term being applied to the complete epileptic seizure, the latter to the passing symptoms of vertigo, headache, brief spasmodic affection of individual muscles, and the like, which frequently occur in the intervals, but are not accompanied by the entire loss of consciousness or the general convulsions that characterize epilepsy in the accepted sense of the term. Epileptic vertigo, like epileptic headache, often exists for a long time before the outbreak of the epileptic paroxysm. Thus one of my patients was for three years subject to giddiness, accompanied by sickness, but without having any confirmed fits. Five years after the vertigo returned with headache, and was now accompanied by entire loss of consciousness. Another for three or four years occasionally felt giddy, and for a short time lost the control over her speech, but without ever becoming unconscious, when she suddenly was seized with an undoubted epileptic paroxysm, in which she struck her face and breast, and bit her tongue severely.

If we are consulted on cases of this kind before the actual epileptic attack has occurred, we may have a doubt as to the head under which it should be classified, but it is impossible to see a patient under the influence of one of these brief attacks without recognising its nature. One of my patients, a surgeon's son, aged fourteen, had at the time of my seeing him been subject to slight epileptoid seizures for four or five years, without ever having a full attack; they recurred several times in one day, and consisted in brief unconsciousness without spasm or collapse; he winked with his eyes in a peculiar manner, having no premonitory symptoms or after feelings beyond the consciousness of a chasm in his memory. The tonicity of the muscles was not lost during the seizure, for he remained sitting on his chair, as I witnessed in my own room. Another patient, a young married lady, subsequently subject to severe epilepsy, for some years previously was liable to frequent slight twitchings of one eye and the side of the face, which I have never seen, but which from the description were evidently of the character of an epileptiform attack.

Again, I find an attack of *petit mal* affecting a patient who was subject to long-standing epilepsy, recurring in the complete form about once a fortnight, thus described from observation in my notes:—
"During the consultation, she suddenly stared at me, lost her consciousness without screaming or falling off the chair; began violently to open and shut her eyes, which I could not arrest; this lasted about five minutes; during the attack she wanted to put her fingers into her mouth. The pulse was very small at

the time. On recovery, she could scarcely see. These minor attacks occur ten and twenty times in the day."

On the other hand, some patients never have anything resembling the *petit mal*, while again I have on record others in whom a sort of alternation is observed between the *petit* and the *grand mal*; when the former predominates the latter is in abeyance, and *vice versâ*, as if a multitude of small discharges were equivalent to a single great one.

In some patients the brief attacks of vertigo and semi-unconsciousness, like a cloud passing over the mental horizon, are very frequent; the more often they occur, the more the issue is to be feared, for they appear to indicate a lower tonicity or tension of the nervous system, and a more complete subjection of the individual to the morbid influence.

Thus, to take a single instance, out of many—a boy whom I am in the habit of seeing frequently, and who at long intervals is subject to very severe epileptic attacks of unusual duration, during the free intervals, when apparently in good health, will suddenly cling to his nurse, or anybody who happens to be near him, and cry out, "I'm unwell, I'm unwell!" at the same time that a deadly pallor overspreads his face. This condition is almost instantaneous, and recovering himself, he reassures the bystander by saying, "I am well!"

The memory fails in proportion as these attacks occur; articulation becomes impeded, dysphagia is at times observed, there is a difficulty in retaining the saliva in the mouth, hence dribbling results, and causes the boy and the girl when approaching puberty, or the adult man and woman, again to put on the semblance of infancy.

Doctors Foville and Copland are at issue with reference to the influence exerted by the epileptic vertigo on the intellect; the former asserting that it brings on intellectual decay more frequently than the severe fits; while the latter maintains that "the more severe the fits, the more is this result to be dreaded." My experience leads me to coincide with Foville and with Esquirol upon this point. The latter, in speaking of the influence of epilepsy in causing mental derangement, states that this tendency to dementia bears "a more direct ratio to the frequency of the vertiginous attacks than to that of the epileptic seizures; the vertigo exerts a more active, a more energetic influence upon the brain than what is called the *grand mal*, or the complete fit. The vertiginous attacks destroy the intellect more rapidly and more certainly, although their duration may be almost inappreciable; because there are individuals who may be vertiginous in the presence of other parties, without their being able to perceive it, unless previously informed."

It has appeared to me that the vertiginous attacks (the *petit mal*) come on without the same necessity for a definite exciting cause, such as may be commonly traced antecedent to a complete seizure; and that the former indicate a more decided and persistent debility or lesion of the cerebrum than exists in simple epilepsy. To my mind it justifies the suspicion that the cerebral lobes have become implicated in a manner and to a degree which, I apprehend, does not prevail in that form of epilepsy which is unaccompanied by vertiginous attacks.

These observations lead to the consideration of a subject which, legitimately, ought perhaps to be postponed until we have examined into the causes and pathology of epilepsy; but the circumstances just alluded to have so direct a bearing upon the general effects produced by epilepsy, that I propose at once to discuss the whole question of the sequelæ resulting from the disease. A single fit, as I have had occasion to remark, may recur, never to return, and without leaving any trace of disease. More commonly, however, unless the disease be arrested and the habit broken, the fits recur with gradually increasing frequency, and it is then that we soon discover that the intellectual faculties begin to fail. The memory is the one which first shows impairment; it becomes less precise and tenacious. The reasoning powers yield in course of time to the debilitating influences that destroy the healthy activity of the cerebral lobes; we thus find the confirmed epileptic presenting every stage of mental infirmity, from a mere shade of intellectual inferiority to drivelling idiocy. This is the more painful because, in so many instances, the most hopeful and promising children become the subjects of this malady. The patients themselves often complain that they can no longer retain their recollection of past events; circumstances at times that happened the same day, and the day before, are more difficult to be remembered than others of a much earlier date; thus, one man stated that he could scarcely recollect anything of the previous day, but remembered things that happened "years back" much better. The same individual (in whom the disease was diagnosed as being brought on by an exostosis, or a similar condition of the interior of the skull, resulting from a fall) complained of often feeling a numbness in his fingers.

A temporary paralysis of a part, or even of the whole body, but then with a predominance on one side or the other, not unfrequently remains after the fit. This may happen once and not return. Thus, in one of my cases, the whole of one side was paralysed after fits that occurred three years previous to my seeing the patient; but, although the fits recurred, the paralysis did not again show itself in this form. In this case, however, the paralysing influence of the disease is manifested in the fact, that ever since the patient was first seized he has often stammered; he is also found often to walk and talk in his sleep.* In others the paralytic condition becomes more or less permanent, and is temporarily aggravated when the fit comes on. Or we find a paralytic condition of certain muscles of an extremity, and a permanently excited state of certain other muscles, inducing considerable deformity of the part. In this way distortions of the feet and hands are produced, which the patients carry to their graves. Pes equinus and other forms of club-foot may result from the epileptic paroxysm, though fortunately the complication is not very frequently met with. In a girl of fourteen years, who had been subject to fits since her sixth year, and continued liable to them in an aggravated form at the time I was consulted, the right hand presented a peculiar contraction, which came on after venesection had been employed for the relief of a paroxysm. The hand was partly flexed at the wrist; the fingers and thumb being extended, and the fingers somewhat drawn back towards the dorsum of the hand, so as to form a hollow at the metacarpus; the joints had not, as is sometimes the case, become anchylosed, but the parts could be restored to their

---

* The relation between somnambulism and allied states and epilepsy was long since pointed out.

normal position; yet, on being set free, at once returned to their abnormal state.*

A similar condition prevailed in the case of a lady whom I saw with Mr. Verral some three years ago, and who was subject to epilepsy from the period of dentition; at the first attack the right side was paralysed, leaving sensation unimpaired. When I saw her she was twenty-four years of age; the whole right side was somewhat atrophied, but though the lower extremity had by surgical appliances almost recovered the normal condition, there still was considerable contraction of the hand and arm; this could easily be overcome by force, but on relaxing the effort, the fingers instantly resumed their bird-claw position. I may mention also that this case seemed an instance justifying an attempt at treatment even in apparently hopeless cases; the lady had her paroxysm almost nightly; a free interval never extending beyond a day or two. Still the exhibition of valerianates of iron and zinc, combined with appropriate laxatives and a tonic regimen, caused a postponement of the seizures for an entire month. As the lady resided at a great distance from London, I have not again been in communication with her, and am not aware of the ultimate issue.

This case suggests the consideration of a topic of particular practical interest to the surgeon and physician, the relation between spasmodic diseases and the distortions and deformities resulting from tonic contractions of the muscles. It is known that the latter may be congenital or acquired, and in both these cases

---

* The reader will find an instructive chapter on the relation of paralysis and epilepsy in Dr. Todd's Clinical Lectures on Paralysis and Diseases of the Brain, p. 281.

observation shows us that the nervous system and the organs of locomotion are mutually interdependent. During the epileptic paroxysm contraction of the fingers and thumb and of the toes is a 'passing symptom, in which the flexors overbalance the extensors and cause what are technically called carpopedal contractions. In young children these are particularly significant, because they are sometimes seen when there is but little trace of more serious disease; they are often associated with crowing inspiration and other evidences of a spasmodic tendency, so slight as scarcely to attract the attention of the patient's friends; nay, they are sometimes not even regarded with sufficient solicitude by medical men.

Although the carpopedal contractions pass off with the affection of which they are symptomatic, I cannot but regard them as analogues of those permanent spastic conditions of the muscles, congenital or acquired, which commonly fall exclusively under the eye of the surgeon. Mr. William Adams has been kind enough to examine his notes with reference to the point at issue, and informs me that in about one-fourth of the cases of spastic muscular contraction which he has met with, the affection has resulted from infantile convulsive disease. Mr. Adams, assuming a broad distinction to exist between epilepsy and infantile convulsive disease, states that he has seen epilepsy co-existing with spastic muscular contraction in very few cases, and in them he has discovered on inquiry that the muscular affection was not produced by the epilepsy, but had preceded it. " I have found," he writes, "that some difficulty or peculiarity in walking, a liability to fall, a very awkward mode of running, or a difficulty in separating the legs, as in striding a rocking-horse, have indicated the existence of spastic muscular contraction, though in a slight degree, previous to the occurrence of the first epileptic seizure. In one case of epilepsy, co-existing with severe spastic contractions, in a young gentleman of seventeen, now under my observation, both parents considered the epilepsy, which commenced at eight years of age, to be the primary affection; but I learned that the peculiar way in which the boy walked and ran with his legs stiffened, had been noticed some time before the first attack of epilepsy." I think that on a review of the whole subject we may admit a double relation to exist in the case of convulsive disease and permanent muscular contraction; a peculiar excitability of the nervous system is common to both; and as, under certain circumstances, convulsions induce permanent muscular spasm, so occasionally the muscular spasm or any interference with it may at times, by reflex irritation, give rise to the convulsive disorder.

We find, on the one hand, to use Mr. Brodhurst's[*] words, that "children who are born with distortions, such as varus, &c., are unusually subject to convulsive disorders, irritation or excitement of whatever kind being apt to induce epilepsy;" on the other hand, it is observed that epilepsy and epileptiform affections are occasionally followed by a permanent spastic condition of certain muscles, especially of the extremities, rendering surgical interference of some kind necessary. This need not necessarily be by tenotomy, however valuable in the majority of confirmed cases this proceeding is found to be. Mr. Adams, justly a strong advocate of tenotomy in cases of actual deformity, even

---

[*] On the Nature and Treatment of Clubfoot and Analogous Distortions, involving the Tibio-tarsal Articulation, p. 57.

with co-existent epilepsy, urges the employment of frictions, exercise or passive motion, and galvanism, as a means of arresting and quieting muscular spasm, and thus anticipating the more serious and permanent distortion.

Some of the preceding remarks receive illustration from the two following cases with which Mr. Brodhurst has favoured me.

The first is that of a child one year old, operated upon for double varus, which was brought up from the country, where it had lived without other children. At the house to which it was taken in town, the relations by nursing and playing with it excited the child, and after three days it became irritable and pale, and lost its appetite. In the course of a few days " a fainting fit was observed, as the mother said, and on the day following three occurred, with all the characters of epilepsy. From that day the child was kept quiet in its own nursery, and taken out much in the open air; it was not allowed to see the other members of the family. After these precautions were taken the child had no more fits, and it soon recovered."

The second case was one of congenital varus, upon which Mr. Brodhurst operated in February, 1852. During the operation it was seized with an epileptic fit, which caused a brief interruption, after which the tendons were divided. The usual treatment was followed and led to a complete removal of the distortion. " Seventeen months after the operation the child again had an epileptic seizure, when both feet were immediately drawn into the same positions in which they were distorted at birth. The father of this child was epileptic."

This case presents features of considerable interest;

it well illustrates the connexion between epilepsy in the father and congenital clubfoot, while it shows that there is reasonable ground for assuming in the individual the same fundamental origin of both affections. Mr. Brodhurst quotes a case analogous to the last from Guérin's work on the Etiology of Pes Equinus.

Moreau* has analyzed 440 cases of epilepsy in females, which occurred between 1821 and 1851 in the Salpetrière, to determine the relative frequency of paralysis. Of these 80 were paralytic, 9 had been so from birth, 24 were temporarily, 47 permanently affected.

The paralysis
  Preceded the epilepsy in . . 17 cases
  It followed in . . . . . 34 ,,
  It occurred coincidently in . 20 ,,
  The period of its duration
   was unknown in . . . 9 ,,
         ———
    Total 80

The extent to which articulation suffers depends upon two causes—upon the partial paralysis affecting the muscles of the pharynx, larynx, tongue, and face (in fact, the muscles under the domination of the respiratory tract), and upon the extent to which the intellect is impaired. Even in comparative health we often find a hesitating speech brought on temporarily by depressing mental and physical influences; this is much more the case in morbid conditions of the nervous system. Hence in some epileptic subjects we find a permanent impairment of the speech induced by a

---

* De l'Etiologie de l'Epilepsie. Mémoires de l'Académie dé Médecine, tome xviii.

paralysis of certain muscles, as in the case of the distortions of the extremities just spoken of; in others, an occasional impairment, or a temporary aggravation of the difficulty of articulation, results from the general depression of the nervous system induced by repeated paroxysms.

In extreme cases there is dribbling at the mouth from partial insensibility to the stimulus of the saliva and the constant half-closed state of the lips. The tongue appears to move with difficulty, and deglutition is not effected with the ordinary ease, not so much from any actual paralysis of the muscles involved in this act, as from the feeble state of the will and the want of power in co-ordinating the complex movements necessary for the process.

Associated with the difficulty of articulation is a prevailing expression of hebetude; in an extreme degree, amounting to idiocy, but in minor degrees characterized by a peculiar heaviness about the eyes, a pasty, leaden, or livid hue, a thickness and coarseness of the lips, which the experienced eye will not fail to recognise. There is even something in the gait of a confirmed epileptic which a close observer will note as indicative of the malady. The cutaneous circulation seems to be no longer carried on in channels possessing vitality, but to flow sluggishly through uncontractile tubes. A similar sluggishness becomes the type of the animal and organic functions of the confirmed epileptic; the circulation of the surface being particularly feeble, epileptics are chilly and liable to coldness of the extremities; thus one patient writes from the country that, since being affected with the disease, she is subject to very cold feet. Another tells me that he is troubled with cold hands, which become discoloured after the fits, possibly, as Dr. Marshall Hall would suggest, from compression of the subclavian vein.

As we find that epilepsy occurring early in life is the most curable, so, on the other hand, when persistent, it more frequently and speedily induces mental derangement, characterized by imbecility. There is some difficulty in obtaining satisfactory statistics on this point as on some others relating to epilepsy, because confirmed epileptics are so frequently removed from the observation of the physician who saw the commencement of the disease, to be placed in asylums; hence the statistics of these establishments only refer to a certain portion, but by no means to all epileptics, as they exclude nearly all cases that have been cured, or in whom the disease has not reached a maximum of intensity. To the comprehension of the disease these cases are as important as, and perhaps even more so than, the instances in which art is, at present at least, unable to achieve any satisfactory result. With this warning, I quote the numbers given by Esquirol.*

Of 385 epileptic females in the female department of Charenton, 46 were hysterical; of the remaining 339,—

    12 were monomaniacs,
    30 „ maniacs,
    34 „ furious,
  145 „ demented,
    8 „ idiots,
  50 „ habitually reasonable, but afflicted with frequent loss of memory, and
  60 exhibited no aberration of intelligence.

---

* Maladies Mentales, tome i. p. 274, seqq. Paris, 1838.

SEQUELÆ OF EPILEPSY.

Hence four-fifths were more or less deranged in their mind; one-fifth preserved their reason; "but," as Esquirol adds, "what reason!"

As mind and body suffer at all points from a repetition of epileptic seizures, it is not surprising that the duration of life is curtailed in these cases. The fit occasionally proves fatal by the exhaustion of nervous power, or by the interruption of the respiratory process; but more frequently death results from the supervention of other diseases, or from the complications with which epilepsy is associated. The frequency of death from epilepsy bears no proportion, it appears to me, to the frequency and the importance of the disease itself; a source of some comfort, however slight, to the patient and the patient's friends, as all people have a fear of a death under such circumstances. From the Registrar-General's returns I have calculated that for 1850, 1851, 1852, and 1853, the mortality in England from epilepsy was, respectively, 0·44 per cent., 0·44 per cent., 0·47 per cent., and 0·50 per cent. of the total mortality; the number of deaths being in—

	1850.	1851.	1852.	1853.
From all causes	368,602	395,396	407,135	421,097
From epilepsy	1,630	1,760	1,935	2,120

These numbers would give an average of 1,861 deaths annually from epilepsy with a ratio of 0·46 to the deaths from all causes. As we have no positive knowledge of the numerical frequency of the disease, the information obtained by the returns of the Registrar-General can only be regarded as an approximative indication. There can scarcely be any difference of opinion as to epilepsy being very much more frequent than these numbers would imply, if we judged of the frequency of the disease by its apparent fatality.

If we turn to page 96, we find good evidence of the fact, that there are about 18,000 male adult epileptics in England; it is not an unreasonable estimate to allow the same number for children and for adult females, which would bring the total number of epileptics in this country to about 54,000 in round numbers. Any one who will examine the basis of my calculations will, I think, be disposed to admit that this is not an excessive estimate.

The comparative table of deaths from epilepsy and from all causes, for London only, extracted from the Registrar-General's "Sixteenth Annual Report," on the next page, yields similar results as the former numbers.

In the tables which can be constructed from the Registrar-General's reports, it is necessary to remember that there must be certain sources of error, owing to the uncertainty of the returns, and certain objectionable features in the nomenclature. The deaths from convulsions, which cause so large a mortality in infancy, are given in the reports under a different head from epilepsy, although in many cases there is not that general and essential distinction which is thereby implied between the two diseases. The addition of even a portion of the cases of convulsions to epilepsy would materially alter the ratio of epileptic mortality in early life.

Thus we find the deaths throughout England in 1853 to have been—

	Males.	Females.
All causes	214,720	206,377
Epilepsy	1,158	962
Convulsions	13,977	10,819

If we compare these numbers with the table on the opposite page, we at once perceive either that epilepsy

MORTALITY FROM EPILEPSY.

No. 3.

		All ages.	Under 1 year.	1.	2.	3.	4.	Under 5 years.	5.	10.	15.	25.	35.	45.	55.	65.	75.	85.	95 and upwards.	Not specified.
Deaths from all causes in London in 1853	M.	30,852	7,302	2931	1535	881	540	13,245	1172	522	1683	2233	2603	2689	2544	2466	1359	290	19	37
	F.	28,217	5,679	2886	1503	891	509	11,468	1140	479	1666	2051	2279	2257	2460	2796	2094	568	40	19
	Total	60,069	12,981	5867	3038	1772	1055	24,713	2312	1001	3249	4284	4882	4946	5004	5262	3453	848	59	56
Deaths from epilepsy in London in 1853	M.	202	5	5	1	4	—	15	7	7	18	31	44	30	16	15	13	—		
	F.	183	1	2	1	2	2	8	3	6	24	25	28	32	23	22	11	1		
	Total	385	6	7	2	6	2	23	10	13	42	56	72	68	39	37	24	1		

The per-centage of deaths from epilepsy at all ages as compared with the deaths from all causes is, according to this table, 0·64, or somewhat higher than the mortality for all England, calculated previously.

and convulsions must be regarded as essentially different, or that the nomenclature is ill-chosen; for the mortality at the different ages pursues a totally different ratio in the two cases. In the case of epilepsy the deaths occur chiefly after puberty; in that of convulsions, they almost exclusively affect the first years of life.

Thus, on calculating the per-centage of mortality from all deaths under one year for 1853, we find it to amount to 25·5, or above one quarter. The average number of deaths occurring at the same period of life from epilepsy annually, is 0·4; whereas we find that, of the 2162 deaths set down to convulsions for London in 1853, 78·4 per cent. occurred during the first year of life alone, and 98·5 per cent. during the first five years (see Table No. 4). This I deem conclusive that there is a necessity for the revision of the nomenclature, if it is not intended to mislead.

Correctly to estimate the influence of epilepsy upon the duration of life, we ought to know the average duration of the disease after it has once shown itself; for this, too, there are no data even of an approximative character, so multiform is the disease; moreover, on comparing the two mortalities for England and for London, we observe that the greatest mortality from epilepsy in the former case falls in the decennial period from 15 to 25, in the latter from 35 to 45. The greatest general mortality in England and Wales, after the first ten years of life, occurs between 65 and 75; the greatest general mortality for London, in males, however, it is to be observed, takes place in the decennial period 45—55, which must not be overlooked in our statistics of epilepsy; since we shall find, in examining into the proclivity of the two sexes to the disease, that

No. 4.

Deaths in London, at different Periods of Life, in 1853.

		Total.	Under 1 year.	1.	2.	3.	4.	Under 5 years.	5.	10.	15.	25.	35.	45.	55.	65.	75.	85.	95 and upwards.	Not specified.
All causes	Males	30,852	7302	2981	1535	881	546	13,245	1172	522	1683	2233	2603	2689	2544	2466	1359	280	19	37
	Fem.	29,217	5679	2886	1503	891	509	11,468	1140	479	1566	2051	2279	2257	2460	2796	2094	568	40	19
Epilepsy	Males	202	5	5	1	4	—	15	7	7	18	31	44	36	16	15	13	—	—	—
	Fem.	183	1	2	1	2	2	8	3	6	24	25	28	32	23	22	11	1	—	—
Convulsions	Males	1,243	965	154	48	28	13	1,228	9	1	2	1	—	1	—	—	—	—	—	—
	Fem.	920	712	118	48	16	9	903	12	2	1	—	—	1	—	1	—	—	—	—

it preponderates, in England at least, in an undoubted manner, in males.

Of the immediate cause of death in epilepsy we have no data. The lesions associated with the affection will be spoken of under the head of morbid anatomy; but of the fatal symptoms which are at times seen, we must advert to those of an apoplectic character. Apoplexy, as we have already observed, is not always easily distinguished from epilepsy; in fact, the two diseases frequently pass into one another. Prichard* specially dwells upon the relation of the two diseases.†

---

* A Treatise on Diseases of the Nervous System, part i. pp. 85, seqq.

† See also Copland's Dictionary, vol. i. p. 795.

CHAPTER V.

The cause of epilepsy — The demoniac controversy — Dr. Adams' opinion — Dr. Bucknill's views — Opinions of Churchmen — The frequency of epilepsy in England; in different troops, in the whole population, in France — Dr. Boudin's statistics — Epidemics of epilepsy — Sex — Age — Hereditary influences — State of the organs of excretion and secretion — The kidneys — The digestive organs — Costiveness — The sexual organs — Marriage — Masturbation — Sexual derangement in males and females — Should epileptics marry?

WE now approach the debateable ground of the causes of epilepsy. They have been classed under different heads by all authors who have ever written upon the subject, from the time of Galen downwards.

This would be an appropriate place to discuss the question of demoniac influences in the production of epilepsy, as one of the solutions of the difficulties that present themselves, to which, in the earliest periods of history, attention was directed, did the subject not tend to lead us into the dangerous field of theological discussion. Although it would be incompatible with the object of this volume to pursue a detailed investigation into this matter, a few remarks showing the two main aspects of the inquiry, as they have been regarded by different minds in all ages down to the present time, may not be out of place.

There have ever been two distinct modes of viewing the relation of the immaterial and the material world. The antagonism between those who believe in a constant and direct manifestation of the spiritual world to our senses; and those who, while believing in spiritual influences, dispute the force of the evidence that is brought to prove their palpable character; has ever existed, and will ever continue to exist. It was manifested in regard to epilepsy at the time of Hippocrates, who eloquently and energetically combated the demoniac origin of the " sacred" disease. " The disease called sacred," he says, " arises from causes like the others—namely, those things which enter and quit the body, such as cold, the sun, and the winds, which are ever changing, and are never at rest; and these things are divine, so that there is no necessity for making a distinction and holding this disease to be more divine than the others; but all are divine, and all human. And each has its own peculiar nature and power, and none is of an ambiguous nature, or irremediable."* Still all through ancient times, again in the Middle Ages, and even in our own time, demoniac possession has had and has its supporters. Dr. Bucknill, in his valuable and interesting work, entitled "The Psychology of Shakespeare,"† in speaking of the scene between the clown, as Sir Topas the curate, and Malvolio, in *Twelfth Night*, observes that it is intended as a caricature of the idea that madness is occasioned by demoniacal possession, and is curable by priestly exorcism. "The idea," he says, "was not merely a vulgar one in Shakespeare's times, but was maintained long afterwards by the learned and pious; more than a trace of it, indeed, remains to the present day in Canon lxxii.

---

* The Sacred Disease. Dr. Adams' translation of the Genuine Works of Hippocrates, vol. ii. p. 857.
† London, 1859, p. 256, seqq.

of the Church, which provides that no minister without the licence of the bishop of the diocese shall 'attempt, upon any pretence whatsoever, either of possession or obsession, by fasting and prayer, to cast out any devil or devils, under pain of the imputation of imposture, or cosenage, or deposition from the ministry.'"

Dr. Bucknill goes on to say that he has known the ceremonial of exorcism performed more than once without reference to episcopal authority, which, he thinks, was intended to check injudicious zeal in the employment of a superstitious rite. Dr. Bucknill holds that the ceremonial of exorcism was of no uncommon occurrence in Shakespeare's time. He states that in Catholic countries it is still resorted to, and that in the lunatic colony of Gheel, in Belgium,* it appears to be the usual active treatment to which recently admitted patients are subjected.

Among less civilized nations the belief in demoniac

---

* In Dr. Webster's instructive and interesting Notes on Belgian Lunatic Asylums (see Winslow's *Journal of Psychological Medicine*, 1857, p. 210), we find a full account of the origin of St. Dympna's shrine at Gheel, where the exorcisms are practised, and of the character of the performances; the sublime and the ridiculous would appear in close relation, did not the whole impress one as a painful memorial of human folly. The legend of St. Dympna is to the effect that, having offended her father, an Irish king, from whom she fled to Belgium, she was pursued by her parent, who struck off her head with his sword. Several lunatics were present who were cured by the horrible spectacle, hence the place acquired a reputation for the cure of insane persons. It is only just to add, that Dr. Webster states the superstitious practice of exorcism to have almost fallen into desuetude—a view which is supported by the fact that six eminent physicians and surgeons have the medical superintendence of Gheel.

possession continues almost universal. For instance, in the "Voyage en Abyssinie, par Lefebvre, Petit, Quartin Dillon" (Paris, 1844), we are told that the most deep-rooted superstition prevailing among the Abyssinians is that in *boudas,* or sorcerers, whose business is to exorcise the demons, to whose vagaries diseases, especially of an epileptiform character, are ascribed.*

Dr. Adams, in a note to another passage of Hippocrates, adverts to the demoniac possession so frequently spoken of in the New Testament, and expresses himself strongly in favour of the opinion that the term δαιμονιζόμενοι (" the possessed with devils") of Sacred Writ, was only employed in compliance with the prevailing form of expression. "That the persons thus described," he adds, " as being possessed with impure spirits were the same as the demoniacs of the Greeks, and that they were epileptics and maniacs, cannot admit of the very slightest doubt." A similar view is maintained by Dr. Cooke,† who enters fully into the inquiry. Even Mead, who cannot be accused of a want of sympathy with the supernatural, says in his "Medica Sacra" (p. 83) of gospel lunatics, " Ii autem aut insani erant, aut insani simul et epileptici."

Among theological writers, we find a great diversity of opinion on this interesting topic. Among those who adopt the same line of argument as the medical writers just quoted, may be mentioned Farmer,‡ Light-

---

* See also on the subject of Tigretier, Hecker's Epidemics, translated by B. G. Babington, M.D., F.R.S., p. 133. Syd. Soc. 1844.

† A Treatise on Nervous Diseases, by John Cooke, M.D., F.R.S., vol. ii. part 2. London, 1823.

‡ Essay on the Dæmoniacs of the New Testament. London, 1775. And Answer to Dr. Worthington on the same, by Rev. Hugh Farmer.

foot,* Wright†, and Lardner.‡ The first is the most interesting and comprehensive of these authors, and brings forward a large body of evidence from Homer and Herodotus downwards, to show that diseases were ascribed by Jews and Gentiles to spirits, and argues that the sacred writers use the phrase *having a demon* in the same sense as their predecessors and contemporaries did—namely, as applied to persons of unsound understanding.

The other side, the literal interpretation of the New Testament term demoniac possession, is upheld by Warburton,§ Jortin,‖ Bloomfield,¶ Worthington,** Fell,†† and numerous others. I quote a few remarks from Bloomfield as more accessible than the other works, though what he says is mainly a *réchauffé* from Warburton. While admitting that there are some passages in Scripture which prove that the terms σεληνιαζόμενοι, ἐπιληπτικοί and δαιμονιζόμενοι, were sometimes used synonymously, he considers that generally they could not be taken as applicable to the same disorders, because those

---

* Horæ Hebraicæ et Talmudicæ, by John Lightfoot, D.D., edited by the Rev. Rob. Gandell, M.A. Oxford, 1859. See vol. ii. p. 248.

† Essay on the Existence of the Devil, by Richard Wright. 1810.

‡ Lardner's "Remarks on Dr. Ward's Dissertation," in his Works, vol. x. p. 265; and in his work on the Credibility of the Gospels.

§ Divine Legation, lib. ix. and sermo xxvii.

‖ Ecclesiastical History, i. 268.

¶ The Greek Testament, with Notes, by the Rev. S. T. Bloomfield. Second edition, 1836, vol. i. p. 20.

** Worthington: Impartial Inquiry into the Case of the Gospel Dæmoniacs, &c. London, 1777.

†† Fell on Dæmoniacs. London, 1779.

possessed with demons were precisely *distinguished* not only from natural diseases of the worst sort, but from lunacy in particular. Dr. Bloomfield holds that demoniacal possessions have an intimate relation to the doctrine of *redemption*, and were therefore reasonably to be expected at the promulgation of the Gospel. He considers that the doctrines of *demoniacal possessions* and of a *future state* (the Italics are the Doctor's) were equally supported by the acts and preaching of "Jesus and his disciples, and are equally woven into the substance of the Christian faith ; the doctrines of the *Fall* and of the *Redemption* being the two cardinal hinges on which our holy religion turns."

As the literal version probably has the largest number of advocates, I would draw attention to the circumstance, that frequently the sacred writers use terms implying "possession" in conjunction with other diseases, confessedly of a pathological character, and apply the same terms for the treatment to both; just as we might speak of curing epileptics, phthisical subjects, and lunatics. Thus, for instance, in St. Matthew, iv. 24, it is stated the people brought to Jesus καὶ δαιμονιζομένους καὶ σεληνιαζομένους καὶ παραλυτικοὺς, καὶ ἐθεράπευσεν αὐτοὺς: and again, St. Matthew xv. 22 and 28, ἡ θυγάτηρ μου κακῶς δαιμονίζεται—καὶ ἰάθη ἡ θυγάτηρ αὐτῆς. In the first instance the possessed are classed together with other sick persons; and in both the terms applied to the cure of the patients are such as are used for ordinary diseases.

Again, as in Greek and Roman writers generally epileptics and lunatics are synonymous, we find the same in the New Testament and also that here lunatic and demoniac are convertible terms. Thus the same epileptic boy who in Matthew (chap. xvii. v. 15) is called a lunatic, (σεληνιάζεται καὶ κακῶς πάσχει) is spoken of in Luke (chap. ix. v. 38—43) as labouring under demoniacal possession; and it is remarkable that in the latter passage two expressions occur of which the first is commonly allowed not to apply to possession in the sense of which we are speaking; they are (v. 39) καὶ ἰδοὺ πνεῦμα λαμβάνει αὐτόν, and (v. 42) ἔρρηξεν αὐτὸν τὸ δαιμόνιον; and again ἐπετίμησε δὲ ὁ Ἰησοῦς τῷ πνεύματι τῷ ἀκαθάρτῳ.

Whatever interpretation we may be disposed to put upon this and many similar passages, it is impossible not to be struck with the accuracy with which Luke, "the beloved physician," delineates an epileptic seizure in his ninth chapter.*

In whatever light we may regard the interpretation of the phraseology of the New Testament, no one contends that there is any form of disease in the present day which is seriously to be attributed to the influence of evil spirits, unless it be the spirit of gin and brandy. We have, therefore, to deal with agents that are more tangible, and which it is more possible to appreciate in their true bearing upon disease.

---

* For the benefit of those of my readers who may be disposed to follow up this inquiry critically, I subjoin seriatim the passages in the New Testament which have appeared to me to bear upon the whole question :—Matthew, iv. 24; viii. 16; viii. 28, seqq.; ix. 32; x. 1; x. 22, seqq.; xii. 22; xv. 22; xvii. 15. Mark, i. 23, 32, 34; iii. 11, 15; v. 2 to 18; vi. 7, 13; vii. 25, 26; ix. 17, 30, 38 (this passage also very minutely details an epileptic seizure); xvi. 9, 17. Luke iv. 33, 35, 41; vi. 18; vii. 21; viii. 2; ix. 1, 39, 42; x. 17; xi. 14; xiii. 11, 32. John vi. 70; vii. 20; x. 20, 21. Acts v. 16; viii. 7; xvi. 16, 18; xix. 12-16.

In many cases we are able to show the operation of two distinct influences which have co-operated towards the production of the epileptic fit. We have seen that frequently the actual outbreak is preceded by a protracted state of indisposition, accompanied by symptoms which may legitimately be regarded as belonging to the complete picture of the disease, and which indicate a peculiar habit of body. We know of a variety of circumstances, which are found to prevail more or less extensively in epileptics before the disease has manifested itself, and which, like the barrel of gunpowder, require the spark to induce an explosion. The inflammability of different materials, if we may continue the simile, varies much, and in the same way the facility with which epilepsy may be excited in different subjects differs according to their susceptibility. The class of influences which determine and modify this susceptibility are termed predisposing, while those which appear to be immediately connected with, or to stand in the relation of cause and effect to, the outbreak itself, are called exciting influences. It is probable that no first outbreak of epilepsy ever occurs without the concurrence of the predisposing and the exciting influence. We find that all the various circumstances that may be mentioned as inducing a predisposition to epilepsy may affect an individual without causing an attack; and we constantly find that individuals are exposed to the influence of the same circumstances which have been known to excite the paroxysm in others, without themselves becoming epileptic. The complex of symptoms which constitute the fit once having occurred, they may recur under a much more trivial exciting cause than in the first instance; and in many cases it appears as though the first fit was an impulse which had set in motion a series of automatic movements, which returned as often as a certain accumulation of force has taken place, without any other exciting cause than that residing in the natural functions of the body.

The great bulk of the evidence is in favour of the view that the predisposing influences enfeeble the body, and more especially the nervous system. The disease is regarded by the great majority of authors past and present as one of debility and impaired nutrition of an asthenic kind; and the influences that induce it are such as would weaken the individual, and expose him to the reception of noxious influences of all kinds. Hence we must assume something more than the predisposing influences, commonly so called—namely, a peculiar habit of body, which we are certainly unable to define, but which, for want of a better term, may be called a nervous diathesis.

That a material difference in this respect exists in different nations, and in different districts of the same country, appears undoubted. We should expect a different ratio in different classes of society; but we should scarcely be prepared to assume the very marked difference of the liability to the disease between the whole population of two parts of the same country, were we not provided with irrefragable evidence to that effect.

Whether epilepsy be a disease of solitary occurrence or of wide-spread frequency is a point that probably would but little affect the individual sufferer. But to all who take a comprehensive view of disease, and would wish to estimate its bearings upon the general sanitary condition of a race or a nation, it must be a matter of serious import to determine its statistics.

94　EPILEPSY AND EPILEPTIFORM SEIZURES.

Epilepsy appears to belong to all climes and all countries; it occurred in the early history of mankind, and it prevails at the present day,—among the untutored savage as among the most cultivated of civilized society. It startles the mother from the security with which she hangs over her beloved infant; it affrights the lover trusting in the future happiness promised to him by his betrothed; it warns the son and the daughter of the mutability of things when they see a parent, whom they thought healthy, struck down by the convulsive paroxysm. Epilepsy spares no condition, age, or sex, and still there are not many diseases upon which fewer positive statistical records are to be found; a circumstance which may be explained by the fact that it is not itself generally a fatal disease, that the individuals suffering from it are not generally admitted into our hospitals, and that different notions prevail as to what ought and what ought not to receive the name of epilepsy.

If we refer to the Registrar-General's reports,* we find a statement of the number of deaths which are caused by epilepsy, which, when compared with the total deaths of the same respective periods, indicate a very small per-centage of mortality due to this cause. I believe that this must not be taken as an indication of the frequency of the disease, as medical men are so constantly in the habit of meeting with it. Still the statistics that I have been able to compass and which are to be relied upon, offer a remarkable uniformity in their results. I give first a tabular statement of the mortality from epilepsy for London, registered in the

---

* See Sixteenth Annual Report of the Registrar-General of Births, Deaths, and Marriages in England. 1856.

December quarters for the five years 1849 to 1853 inclusive :—

	Deaths from Epilepsy.	Total deaths in same period.	Per-centage of mortality from Epilepsy.
1849	73	12,877	0·56
1850	79	18,544	0·40
1851	75	13,964	0·53
1852	118	13,448	0·87
1853	117	17,390	0·67
Average	92	15,244	0·61

That the per-centage of mortality for London during the period specified is not far removed from the mortality from epilepsy returned for the whole of England is proved by the following table, the materials for which are also drawn from the same work :—

	Deaths from Epilepsy.	Total deaths from all causes.	Per-centage of mortality from Epilepsy.
1850	1630	368,602	0·44
1851	1760	395,396	0·44
1852	1935	407,135	0·47
1853	2120	421,097	0·50
Average	1861	398,057	0·46

The average mortality from epilepsy being in the former case 0·61, in the latter 0·46.

Again, according to another return with which I have been favoured through the courtesy of Dr. Farr, it appears that the total deaths throughout England and Wales from epilepsy during the seven years 1848 to

1854 inclusive, were 12,876, or, on an average, 1839 annually, which is rather lower than the average of the four years 1850 to 1853 given above.

If we take the population of England for any given year, say for 1850, we may easily calculate the average frequency with which epilepsy is a cause of death; and by comparing these statistics with the statistics obtainable from other sources, we may approximatively determine the frequency with which epilepsy attacks the population at large.

The population of England in 1850 was 17,754,000; the deaths from epilepsy in the same year were 1630, or at the rate of 0·009 per cent. of the total population. These numbers, however, receive further confirmation from the statistics of epilepsy occurring in the army which my friend Dr. Balfour has kindly enabled me to present to the reader. It is not to my purpose at present to go into details; it will suffice to say that among troops serving at home and abroad, and whose strength is given for varying periods between 1817 and 1846, amounting in all to 1,061,233, there were 3264 cases of epilepsy with 96 deaths. This gives a mortality of 2·94 per cent. of the deaths from epilepsy, or 0·009 per cent. of the whole strength, the per-centage of the seizures to the whole strength being 0·307. This, it will be perceived, is identically the same result as the per-centage of the mortality from epilepsy relative to the total population of England, which was also 0·009. In round numbers, we may state that 4 out of every 3000 soldiers are epileptic; and if we apply the same data to the total population, we should conclude that the number of male adult epileptics in England in 1850 was 17,754,000 × 0·003, or about 18,000. We are justified in assuming this number to be not far from the mark, because it is proved by my own statistics and the statistics of numerous writers, that the period most favourable for the development of epilepsy is between the tenth and twentieth years, and that the proclivity to the disease diminishes after the latter period.

I have previously suggested that, in order to allow for epileptic women, children, and old people, it would not be unreasonable to treble this amount, which would bring the total number of epileptics in England to about 54,000. We obtain a somewhat higher total if we employ as a basis for our calculation the per-centage of deaths in proportion to the cases of epilepsy, which we have seen to be in the military 2·94. Taking the annual deaths from epilepsy in England at 1839 (the average of 1848-1854) to represent the same number of cases of epileptics as in the military, we obtain a total of 62,551.

If, according to these two estimates, we say that there are on an average about 60,000 epileptics in England and Wales, we shall not be likely to err far.

The statistics for the colonial corps for similar periods as those given above for British troops, yield a higher average of mortality from epilepsy, but a lower frequency of the disease. The total strength of the colonial corps was 129,914; the total cases of epilepsy admitted into hospital 165, and the total deaths from that cause 19; or 11·5 per cent. of the cases of epilepsy were fatal. It is to be observed that 16 of the 19 deaths occurred among Negroes, although they constitute little more than half the entire force.

On calculating the ratio of deaths from epilepsy to the total strength, we find it to be 0·014; while 0·12 per cent. of the total strength was affected with epilepsy.

The results obtained by the analysis of the statistics of the colonial corps differ considerably from those yielded by the British troops; the number of deaths from epilepsy in the former being proportionately very much larger, while the seizures are more numerous among the British troops.

Comparative Statistics of Epilepsy.

	Total strength.	Number of cases of Epilepsy.	Number of deaths.	Percentage of deaths to cases of Epilepsy.	Percentage of seizures to total strength.	Percentage of deaths to total strength.
British troops ...	1,061,233	3264	96	2.94	0·307	0·009
Colonial corps ...	129,814	165	19	11·2	0·12	0·014
Population of England in 1850	17,754,000	unknown	1630	......	......	0·009
French recruits (in 23 years)	4,036,372	6627	......	......	0·16	

We shall feel less surprise at the marked difference presented by the two classes of troops in her Majesty's service when we examine the returns which the French system of conscription places at our disposal. In the following table is given the number of young men of twenty-one years of age who were exempted from military service by reason of their liability to epilepsy, the fact being proved by the certificate of " three fathers of families residing in the same canton."* The variations in the frequency of epilepsy range from 41·5 to 339·0 per 100,000 recruits :—

---

* It is not stated that a medical certificate is required, but we may assume, that where conscription is an essential part of the administration, no extensive system of fraud can well be carried out.

FRENCH STATISTICS.

Department.		Department.	
1. Puy de Dôme	41·5	40. Dordogne	144·1
2. Manche	60·2	41. Corse	145·5
3. Haute Vienne	76·2	42. Arsne	150·3
4. Loiret	78·4	43. Allier	150·4
5. Seine et Marne	82·1	44. Pas de Calais	153·4
6. Yonne	82·6	45. Nord	155·3
7. Tarne et Garonne	85·9	46. Basses Alpes	156·0
8. Aude	86 8	47. Arveyron	158·1
9. Indre	87·6	48. Gironde	158·3
10. Rhone	88·5	49. Vaucluse	160·2
11. Meurthe	93·5	50. Nièvre	163·7
12. Côte d'Or	} 93·9	51. Maine et Loire	166·6
13. Doubs		52. Haute Saône	169·7
14. Deux Sèvres	98·3	53. Vienne	174·2
15. Finistère	100·5	54. Ile et Vilaine	178·6
16. Ain	105·9	55. Seine et Oise	183·5
17. Bas Rhin	106·8	56. Oise	184·3
18. Vosges	109·8	57. Lot et Garonne	184·4
19. Calvados	111·5	58. Eure et Loire	185·7
20. Lot	113·7	59. Drôme	186·9
21. Ardennes	117·9	60. Indre et Loire	187·3
22. Jura	118·5	61. Hautes Pyrénées	188·2
23. Cantal	120·7	62. Loire et Cher	194·7
24. Tarn	123·7	63. Hérault	196·4
25. Saône et Loire	124·4	64. Landes	197·8
26. Moselle	125·3	65. Isère	201·8
27. Hautes Alpes	127·7	66. Gers	202·8
28. Charente	} 130·9	67. Morbihan	203·1
29. Orne		68. Sarthe	204·2
30. Charente Inférieure	131·1	69. Haute Marne	204·4
31. Côtes du Nord	132·2	70. Haute Loire	209·1
32. Eure	} 133·4	71. Var	210·3
33. Gard		72. Somme	213·2
34. Ardèche	133·6	73. Haute Garonne	222·4
35. Loire	134·6	74. Mayenne	223·7
36. Seine	137·6	75. Vendée	232·9
37. Creuse	137·7	76. Marne	233·3
38. Haut Rhin	138·0	77. Basses Pyrénées	255·1
39. Cher	139·2	78. Bouches du Rhone	257·1

Department.		Department.	
79. Ariège	258·9	83. Aube	280·5
80. Loire Inférieure	261·2	84. Corrège	285·5
81. Seine Inférieure	274·2	85. Meuse	296·3
82. Lozère	277·2	86. Pyrénées Orientales	339·9

In the important work of M. Boudin,* from which the foregoing table is extracted, we find other interesting facts bearing upon the frequency of epilepsy. "From 1831 to 1853 inclusive, a period of twenty-three years, there were 6627 exemptions on the score of epilepsy among 4,036,372 young men examined by the *Conseils de Révision*, or at the rate of 164 per 100,000." This is equal to 0·16 per cent.; a lower amount than met with in our troops, which may be accounted for by the greater absence of exciting causes among recruits than among soldiers in active service, as well as by the fact that all epileptics are rigorously excluded from the service. "During the same periods," M. Boudin continues, "the proportion of exemptions to 100,000 men examined varied as follows:—

Year.	Numbers exempt.	Year.	Numbers exempt.
1831	269	1843	153
1832	220	1844	163
1833	198	1845	141
1834	178	1846	147
1835	159	1847	173
1836	151	1848	170
1837	154	1849	167
1838	169	1850	141
1839	168	1851	140
1840	160	1852	124
1841	143	1853	135
1842	155		

* Traité de Géographie et Statistique Médicales et des Maladies Endémiques. Par J. C. M. Boudin, Médecin-en-Chef de l'Hôpital Militaire du Roule. Tome ii. pp. 449, seqq. Paris, 1857.

"We should not," M. Boudin says, "at once conclude from the preceding table that the number of epileptics has diminished in France since 1831. It would probably be more just to conclude that since 1834, and especially since 1833, the recruiting system has been carried out with more justice. It is certainly worthy of remark that since this period the proportion of exemptions from epilepsy has offered a great uniformity. However that may be, if we consider that the average number of young men of twenty-one is from 300,000 to 310,000, we may conclude, from the annual average of exemptions by reason of epilepsy (164 in 100,000) that there are about (3×164) 492 young men of twenty-one years of age in France who are epileptic."

I have met with no satisfactory statements in authors regarding the influence of race upon the production and the course of epilepsy. But although, at first sight, on comparing the statistics of British troops with colonial corps, there would appear a ground for the inference that a marked difference exists, we are warned to be careful of jumping at any conclusion by the table of the French departments, which shows a difference varying from 41·5 to 339·9, under circumstances which are not, in my opinion, to be accounted for by any difference in race.

M. Boudin does not enable me to solve the difficulty; I shall therefore avoid adding to it by offering an hypothetical suggestion; but it will be well to extract the few remarks which the author referred to himself makes upon the question. "May we," he says, "conclude that epilepsy is in certain cases an endemic affection, properly so called? We do not think that our documents justify this deduction, and it is not impossible that the extreme relative frequency of epilepsy in certain departments is due to hereditary influences, possibly even to race; still we are far from wishing to deny absolutely the possibility of an endemic influence. Some authors have thought that epilepsy was much more frequent in mountainous districts than in the plain. The preceding table proves just the reverse. For the minimum of epilepsy is found in the Puy de Dôme, while the department of the Bouches du Rhone, which is but very slightly mountainous, presents 257 annual exceptions among 100,000 recruits."

If we seek to compare the results deducible from M. Boudin's table with those yielded by the comparison of British and colonial troops, we have no standard of measurement, and the difficulties which M. Boudin experienced in determining the causes of variation in different parts of the population of his own country, become still more manifest when we have to deal with men so little amenable to scientific inquiry as the members of the colonial corps. What the determining circumstances may be which influence the prevalence of epilepsy in certain races, remains to be discovered. Further statistics, too, of a reliable character, would be desirable; but until they are obtained, an inquiry into the hygienic conditions of some of the departments of France, for which we possess satisfactory statistics on this point, would promise valuable conclusions.

The preceding observations prove that epilepsy may be regarded as a sporadic disease, favoured and promoted by certain endemic causes, to which we at present, however, possess no further clue than the evidence of their existence.

Dr. Guggenbühl, the well-known founder of the Abendberg, in a recent work on Cretinism,* makes

---

* Die Erforschung des Cretinismus und Blödsinns nach dem jetzigen Zustande der Naturwissenschaften. Wien, 1860.

some remarks which may tend to assist in the determination of the nature of these endemic influences; he has found that the stimulating air of the Abendberg in winter aggravates epileptic attacks, which are ameliorated in the same locality in summer.

At times, however, the disease has made its appearance in a distinctly epidemic form. Such was the case at the time of the dancing mania which afflicted the people of Aix-la-Chapelle, Cologne, Strasburg, and many of the Belgian towns, in the second half of the fourteenth century. Epileptiform convulsions formed the commencement of the disease, when the affection was completely developed. "Those affected fell to the ground senseless, panting, and labouring for breath; they foamed at the mouth, and suddenly springing up, began their dances among strange contortions."*

A more defined epidemic of epilepsy has occurred during our own time, and has been described and watched by a British physician.† The epidemic affected the inhabitants of Teheran in 1842; it was confined to the months of January and February, and the fits as described appear to have been undoubted epilepsy. Dr. Bell, who was an eye-witness, gives the following delineation of the fit: "Powerful convulsions of one side; for a short time quite purple in the face and chest. Two or three severe opisthotonic spasms, and horrid grinding of the clenched teeth, as in tetanus. Total insensibility; pulse about 90; very powerfully excited, so much that, although convinced that this

---

* The Epidemics of the Middle Ages. By J. F. C. Hecker, M.D. Translated by Dr. Babington. Syd. Soc. Ed., p. 88.

† Some Account of an Epidemic which prevailed at Teheran in the months of January and February, 1842. By C. W. Bell, M.D., Med.-Chir. Trans., vol. xxvi. p. 223.

disease was purely nervous, and little likely to be benefited by blood-letting, and that when this violent excitement passed it would be succeeded by a condition of proportionate feebleness and prostration, yet I found it absolutely necessary to bleed again, in order to protect the brain." The whole population of Teheran appears at the time to have been affected more or less by the epidemic influence, inasmuch as for some nights they were troubled with a sleeping of the leg and arm of one side. Dr. Bell himself experienced it, and "found the sensation and unnatural excitement of the heart extremely unpleasant; but it ceased after taking a dose of iron."

In schools we occasionally meet with epidemics of epilepsy. The following* is an interesting instance: "The Free School of Bielefeld is a well-aired and not over-crowded room, in which the boys and girls are taught at the same time. A young girl, of the name of Arnold, had for some time been subject to epileptic fits, and had been repeatedly seized during the school-hours, on which account she was forbidden to attend. Apparently restored to health, she was re-admitted, but on the 8th of August, 1837, she was again seized, and was in consequence carried home. A few days afterwards, a strong, healthy girl, who had occasionally accompanied Arnold home, was seized with convulsions in the school-room; on the 14th two other girls, aged respectively twelve and fourteen years, were affected in a like manner, but this did not prevent them making their appearance at school on the following morning. Scarcely, however, had the business of the day com-

---

* *Medicinische Zeitung*, No. 8, 1838, and Forbes's *Medical Review*, vol. vi. p. 531.

EPIDEMICS OF EPILEPSY. 105

menced, when, not only these two, but likewise three other girls, were affected with epileptic convulsions, and the contagion spread with such rapidity, that in less than half an hour above twenty girls were similarly affected. At first the children experienced a feeling of anxiety; they were then observed to grow pale, there was oppression of the chest and the head became affected; trembling of the limbs followed, with loss of consciousness; the thumbs were bent upon the palms, the eyes were distorted, and the patient gave vent to a sudden and anxious cry. The paroxysm in some was of short duration, but in others it continued for hours; excepting Arnold, none of the girls attacked had ever had an epileptic paroxysm. The disease was treated as purely nervous, with valerian, oxide of zinc, indigo, &c., but on the whole with little success; for after the lapse of five months there were very few who could be considered as safe against a relapse." An account of another analogous epidemic may be found in the same place as the one just quoted.

But we have still more recently had something very similar at our own doors. It is impossible to read Archdeacon Stopford's* account of the revivals in Belfast (in 1858-59) without coming to the conclusion that the spurious excitement which accompanied them induced genuine epileptic seizures. Unfortunately there is no accurate report from a medical eye-witness of the phenomena that occurred in these revivals, at least none has come to my knowledge; still the Archdeacon's description is so vivid and minute, that we may fairly

---

* The Work and the Counterwork on the Religious Revival of Belfast, with an Explanation of its Physical Phenomena. By Edward A. Stopford, Archdeacon of Meath. Dublin, 1859.

draw our inferences from it. Many of the symptoms are of an hysterical character, but those of genuine epilepsy are not wanting in some of the cases. An account of another epidemic of an epileptiform affection, termed a convulsive neurosis, is reported by Dr. Eggs,* as having recently occurred in the Normal School of Teachers at Strasburg.†

I proceed to the consideration of the predisposing causes in the individual.

The question of sex and age are those which first suggest themselves. English authors are all but unanimous as to the greater proclivity to epilepsy being on the side of the male sex, while the majority of continental writers take the opposite view. My analysis of one hundred and four cases gives forty-seven females to fifty-seven males, the per-centage being forty-five and fifty-five respectively, omitting decimals. These numbers in themselves would not suffice to determine the question: but, even as far as they go, they prove no marked preponderance on one side or the other. I again have recourse to the statistics of the Registrar-General, which confirm the statements of English authors; but it is to be remembered, that the numbers of deaths by no means represent the numbers of epileptic cases; and that, from the exclusion of almost all cases of "convulsions" in the first five years of life, an error is introduced which I am unable to rectify, as I should as little venture to

---

* Bulletin Général de Thérapeutique, t. lviii. p. 468. May, 1860.

† The celebrated epidemic in Haarlem, which Boerhaave arrested by a threat of the actual cautery, might also be quoted here; but I doubt whether the cases were true epilepsy. (See the report of Boerhaave's nephew in: Impetum faciens dictum, Lugduni Batavorum, 1745, p. 355).

class all convulsions with epilepsy as to regard them all as tetanus.

From the annexed table (No. 5) it would appear that the mortality of males at all ages from epilepsy is 52·26 per cent., of females 47·73 per cent.; and that therefore 4·53 per cent. of male deaths occur from epilepsy in excess of female deaths from that cause; or, to put it in a different way, we find that the average male deaths in one year from epilepsy are 961·3, of female, 878·1; so that annually in England and Wales 83·2 more males die epileptic than females. If only a portion of the deaths from convulsions have to be included in this number, it is probable that this relation would become still more marked; for Dr. Tripe* has shown in an elaborate paper that the deaths of males preponderate over those of females during the first five years of life, from diseases of the nervous system, by as much as 20·5 per cent. Among recent British writers, Drs. Watson and Elliotson may be quoted in support of this view, though they supply no statistics. Romberg maintains the same doctrine.

All French writers regard the numbers indicated by the two great asylums, Bicêtre and Salpêtrière, as conclusive evidence that epilepsy prevails most in females. Esquirol, writing in 1838, states that there were 162 male epileptics at Bicêtre, and 389 female epileptics at Salpêtrière; Moreau, in 1854, gives the numbers respectively at 149 and 234. This would yield a total of 723 females to 311 males, or a ratio of nearly 70 to 30. I cannot but think that there must be some fallacy in regarding these numbers as representing the

---

* *British and Foreign Medico-Chirurgical Review*, April, 1857, p. 460.

108 EPILEPSY AND EPILEPTIFORM SEIZURES.

No. 5.

*Deaths at each Age from Epilepsy in England and Wales during the Seven Years 1848 to 1854.**

	Total.	Under 1 year.	1.	2.	3.	4.	Under 5 years.	5.	10.	15.	25.	35.	45.	55.	65.	75.	85.	95 and upwards.	Not specified.
Males	6,727	231	118	91	64	59	563	256	361	996	1063	1039	876	679	559	291	37	2	7
Females ...	6,147	162	90	65	54	58	429	214	329	1174	1054	854	650	600	531	267	42	1	2
Persons ...	12,876	393	208	156	118	117	992	470	690	2170	2117	1893	1526	1279	1090	558	79	3	9

* This table was kindly supplied to the author by Dr. Farr in 1857.

liability of the whole population to the disease; and that Esquirol, in attributing the greater proclivity of females to epilepsy to their greater "impressionability," adopts a conclusion too hastily arrived at.

Age materially influences the occurrence of epilepsy; but in all calculations hitherto made, the majority of convulsive diseases of early childhood have been excluded from the consideration; allowance must be made for the fallacy resulting from this source. I have not ventured to include in my tables some cases of convulsions which I should regard as identical with epilepsy. The following table yields the results of an analysis of one hundred and four of the cases observed by myself:—

Period of first occurrence of fits.	Number of Cases.		
0— 5 6—10 11—15 16—20	19 } 29 10 } 24 } 44 20 }	73 or 70·2 per cent.	From infancy to the age of 20 years inclusive.
21—25 26—30 31—35 36—40	6 } 11 5 } 5 } 7 2 }	18 or 17·3 per cent.	From 21 to 40 years inclusive.
41—45 46—50 51—55 56—68	5 } 9 4 } 1 } 4 3 }	13 or 12·5 per cent.	From 41 to 55 years inclusive.

From this table it would appear that the greatest proclivity to epilepsy is to be found at the period of puberty; a fact upon which there is scarcely any difference of opinion among writers. Nor does the table of the Registrar-General, given at p. 108, militate against it; since epilepsy is very rarely fatal in the first instance, and the greater frequency of its first occurrence before twenty years is perfectly compatible with the greatest fatality of the disease later in life. By reference to the Registrar-General's table, we find that the greatest mortality occurs between fifteen and twenty-five years, but that it is scarcely diminished during the subsequent decennial period, for the ratio in the former one is 16·8, in the latter 16·4 per cent. of epileptic mortality.

Distributed according to sex, and arranged in decennial periods, I find that my cases occupy this order:—

Period.	Males.	Females.
0—10	16	13
11—20	23	21
21—30	7	4
31—40	5	2
41—50	5	4
51—68	1	3
	57	47

The general results yielded by the above tables are confirmed by the data of Moreau,* who favours us with a table in which the ages of 995 epileptics, collected from various French sources, are analysed:—

Epileptic from birth	87
Epileptic in infancy	25
From 2 to 10 years	281
„ 10 „ 20 „	364
„ 20 „ 30 „	111
„ 30 „ 40 „	59
„ 40 „ 50 „	51
„ 50 „ 60 „	13
„ 60 „ 70 „	4

* De l'Etiologie le l'Epilepsie, Mémoires de l'Académie de Médecine, tome xviii.

HEREDITARY INFLUENCES.

This table would seem to confirm my opinion that the occurrence of epilepsy is regulated by the same laws in France which govern it in this country; and that, therefore, the marked difference between the two in point of sex will be shown to depend upon an erroneous basis having been assumed for the calculations.

Hereditary influences are very palpable in epilepsy; they are shown not only by the prevalence of epilepsy in the same family, but also by the co-existence of epilepsy with other nervous affections, and more particularly with mental derangement. Considerable difficulties are always opposed to inquiries into the existence of such affections in families, as there is a great tendency on the part of individuals and their connexions to conceal facts of the kind. Moreover, it is almost impossible to ascertain the state of health of most of the collateral branches of a family, or those removed in the ascending line from the patients—such as uncles or aunts, of grandfathers and grandmothers. And again, although we may succeed in obtaining tolerably complete data, showing the health of the relatives of a large number of epileptics, the value of the results is unsatisfactory, unless we have as a standard of comparison the ratio of the solitary occurrence of epilepsy; that is, unless we know how often epilepsy may occur without being reproduced in relatives. But the fact is, that nervous diseases of all kinds are so widely spread, that but few families entirely escape them; and the more we extend our intimate acquaintance with the domestic relations of our friends and patients, the more frequently do we meet with instances of insanity or epilepsy where we previously imagined that there was complete exemption. I shall not easily forget the startled, the almost guilty look with which a gentleman met my inquiry while I was standing over a son who, for the first time and apparently without cause, was violently convulsed with the epileptic paroxysm. The inquiry was merely whether he had been epileptic himself, and it was made in order to obtain a clue to the attack in question. The father, a hale and vigorous-looking man, shrank from confessing that he himself had been epileptic.

While speaking of hereditary influences it would be wrong not to advert to the occurrence of epilepsy in the child, owing to the pernicious influence exerted upon the mind of the mother. Thus, in one of my cases, the shock produced upon the mother by the sight of a person in epileptic convulsions was accused as the predisposing cause of the subsequent epilepsy in my patient's case. In another, the sudden shock of the suicide of a near relative was supposed to have the same effect. In a third, great anxiety and depression of mind during a first pregnancy, appeared to be reasonably accused of being the predisposing influence of the subsequent epilepsy in the child. The conclusion seems to be as warrantable as the production of epilepsy in one individual by seeing it in another; concerning which I could, from experience, multiply the evidence already quoted from other authors.

I have found that epilepsy prevailed among members of the families of my patients in 13·4 per cent. of my cases; but I have not extended my inquiry as far as Herpin and Moreau have done, who have included every variety of nervous affection among the hereditary influences to which epilepsy might be traced. An hereditary taint of some kind, such as scrofula, cancer, tubercle, and the like, was found much more frequently than the numbers just given would indicate. But such

HEREDITARY INFLUENCES.

an extension is calculated to mislead, because necessarily dealing with vague data; and it would be perfectly reasonable if we go so far to go still farther, and to include other exhausting or debilitating diseases.* However, M. Herpin's† tables are interesting as far as they go, and I reproduce them in the following combination:—

State of Health in 380 Relatives of 68 Epileptics.

Epilepsy	in	10 cases.
Insanity	,,	24 ,,
Suicide	,,	1 ,,
Melancholia	,,	2 ,,
Hypochondriasis	,,	3 ,,
Hysteria	,,	2‡ ,,
Chorea	,,	2 ,,
Sleep-walking	,,	2 ,,
Nervous excitability	,,	3 ,,
Apoplexy	,,	11 ,,
Cerebral softening	,,	1 ,,
General paralysis	,,	2 ,,
Meningitis and chronic hydrocephalus	,,	13 ,,
Mortal Convulsions	,,	1 ,,
Tetanus	,,	1 ,,
		78

It follows that 68 patients had 78 relatives who laboured under some affection of the nervous system;

---

* In the review of Herpin's work on Epilepsy, by Dr. Parkes (*Med.-Chir. Review*, April, 1853), the reader will find some remarks on this question that are much to the point.

† Du Prognostic et du Traitement Curatif de l'Epilepsie, par Th. Herpin, p. 327. Paris, 1852.

‡ I cannot but demur to this number, as it is incredible that not more than 2 relatives of 68 epileptics should have been hysterical.

but of these disorders the author himself holds that only epilepsy and insanity deserve consideration, because he finds that the proportion of their occurrence among the relatives of epileptic patients is so much larger than among the population at large. He calculates that epilepsy occurs between four to five times, and insanity twenty-four times more frequently among the relatives of epileptic patients than in the population generally.

The large number of cases of apoplexy and meningitis among the relatives of his epileptic patients necessarily attracted M. Herpin's attention; he was naturally inclined to assume that they sufficed to establish a definite predisposing relation; but, on comparing them with the calculations of M. Marc d'Espine, made to determine the causes of mortality in the Canton of Geneva, he finds both apoplexy and meningitis occur even more frequently in the population at large than among the relatives of his epileptics.

"The relation of the number of deaths from apoplexy to that of the total deaths (exclusive of still-born children), calculated for eight years, is 40 per 1000. This number must be increased in order to allow for the apoplectics who die of another complaint; if we make an addition of one quarter for this, we obtain 50 per 1000. The relation of apoplexies in the families of our first series is 11 in 243 individuals, or about 45 per 1000. We find, therefore, that the relatives of our patients present fewer cases of this affection than the population at large. . . . . The same calculations applied to tubercular meningitis yield the same conclusion. M. d'Espine finds the ratio to be 38 per 1000. Our first series yields 7 in 243, or 29 per 1000; our second series, 6 in 137, or 44 per 1000; being an average of 36 per 1000."

In seeking for the causes of epilepsy it is necessarily the desire of the medical man to trace it to some local affection, to discover the habitat of the morbid condition. In a disease so manifestly connected with affections of the nervous system, a search into the state of this part of the organism would naturally be made; but hitherto all attempts to demonstrate a lesion by the microscope or test-tube, which should be so constantly associated with epilepsy as to justify our considering it in the light of an essential constituent of the disease, have failed. Palpable as the symptoms of the disease are, no uniform lesion has hitherto been discovered, either during life or after death, by which we could gauge the relation between functional and organic changes as we can in many other diseases. Nor can it be said that epilepsy varies more in its features and circumstances than any other member of the nosological register, concerning the pathology of which we consider ourselves better informed. Still if we have to deal with inflammation of the lung, we know that the deposit of lymph in the air-vesicles, and the objective symptoms, as cough and dyspnœa, bear a certain relation to one another, and we may, as a general rule, measure one by the other; the progress of peritonitis similarly offers a certain ratio between the amount of local changes and the symptoms appreciable by the medical man. In epilepsy neither the nervous, nor the vascular system, nor any of the individual viscera, present lesions in which hitherto a similar uniformity has been noted, though we may now be approaching to a solution of this difficulty.

Of the symptoms referrible to the nervous system we have said enough when discussing the phenomena of the paroxysm. These, and their relation to insanity, show that the chief fault resides in the cerebro-spinal axis. But it is impossible to overlook the manifest relation that the state of nutrition and the blood exercises upon the nervous system in the production of those symptoms; for, of all the influences that we can trace in the production of epilepsy, we see none that operate so frequently as those which are connected with some derangement of nutrition. And yet there is no definite change in the excretions or secretions which can be shown to be a uniform accompaniment of epilepsy. It is here more particularly that we should have hoped for some aid by the modern improvements in analysis, and although something has been gained, much more yet remains to be done.

The only organs of secretion which have been proved to exercise a well-marked influence upon the occurrence of epilepsy, or rather upon epileptiform convulsions, are the kidneys. The bearing of albuminuria upon their production has been frequently observed, since Dr. Bright* first drew attention to the importance of this symptom as an indication of a very serious organic lesion. Sir Henry Halford had previously (1820) pointed out the connexion between apoplectic conditions and suppression of urine, and it appears from the researches of Drs. Prout, Bostock, and Christison, which have since been confirmed by others, that in these cases of imperfect secretion of urine, urea can be detected in the serum of the blood. The epileptic seizures that frequently† occur in parturient women have been

---

* Reports of Med. Cases, vol. ii. p. 446.

† In Dr. Churchill's excellent Manual of Midwifery, third ed., p. 480, an analysis of 190,313 cases of labour, collected from various sources, shows convulsions to have occurred 273 times, or 1·4 per 1000.

shown to be due to various causes which impair the powers of the patient; but in the large majority of them there is albuminous urine, as first shown by Dr. Simpson and Dr. Lever. However, here too, though a predisposing influence, it is evidently not essential, for convulsions occur without albuminuria; and albuminuria, as at other periods, and in the male sex, may occur without giving rise to epilepsy. We are informed by Dr. Churchill that "Dr. Blot found albumen in the urine of 41 pregnant women out of 205, and chiefly in primiparæ; and Dr. Litzmann examined the urine of 131 females, 79 during pregnancy, 80 during labour, and 80 after delivery: albumen was present in 37, and absent in 95; of the 37, 26 were primiparæ." On the other hand, the fact of Dr. Lever having found the urine albuminous in every case of puerperal convulsions but one, which has come under his notice, argues strongly in favour of the view that, however frequent albuminuria may be in pregnancy, there must be even a more intimate relation between convulsions of parturition and this symptom.

No such uniformity is to be traced between the existence of epilepsy and albuminuria in other individuals. Albuminuria when present, by impoverishing the blood, or by the coincident retention of urea in the blood, may, and frequently does, appear to cause epileptic seizures; but in the great majority of cases of epilepsy no palpable derangement of the renal secretion can be detected. Albumen was found permanently present in one of the twenty-three of my cases in which the urine was tested for albumen, temporarily in one; the former proved fatal, and from the state of the kidneys it was manifest that degenerative disease had been going on in them antecedently to the occurrence of the spasmodic action.

Some time back Dr. Goolden* maintained that a saccharine condition of the urine ordinarily accompanied epilepsy.

Since his observations were published I have tested this secretion, either by Moore's, Trommer's, or Barreswil's test, and generally by more than one, for sugar in nearly all the cases of epilepsy that have fallen under my notice; but in none of the many in which the test was applied have I succeeded in obtaining the proper reaction. Our present knowledge of the relation of the kidneys to the system at large, as the chief emunctories, and the bearings of their functional and organic disturbances upon the general health, and especially upon the healthy condition of the central organs of the nervous system, would incline the physician to watch them, especially in a disease like epilepsy; but hitherto no definite ratio has been detected.

In our examinations of the urine of epileptics we meet with every variety of derangement that is found to accompany different forms of dyspeptic conditions.

---

* *Lancet*, 1854, vol. i. p. 656, vol. ii. p. 29. Since writing the above, Dr. Goolden has favoured me with the following communication, which shows that he does not regard saccharine urine as an ordinary accompaniment of epilepsy, but as having been due to a peculiar epidemic influence prevailing at the time at which he met with it: "With regard to the saccharine condition of the urine, I have since found that sugar is a very rare attendant upon epilepsy. I can only account for the nearly constant presence of sugar at the time of my experiments published, by a fact that Dr. Prout observed, which was, that the saccharine diathesis prevailed to a great extent prior to each cholera epidemic, and my experiments were all made under the same condition."

Frequently there is an excess of phosphates; oxalates are often seen; and I have repeatedly found the urine of epileptics exhibit persistently so large a quantity of urea that, on the addition of equal parts of nitric acid, the whole of the liquid was almost solidified by conversion into nitrate of urea. With regard to the excess of phosphates, it indicates, as Dr. Hunt* has suggested, the propriety of administering nitro-muriatic acid; the same applies to oxalates; but my experience does not confirm Dr. Hunt's views as to the frequency of an excess of this particular derangement, or of an accompanying deficiency of urea. I should say that in my cases the specific gravity of the urine has more frequently ranged from 1020 to 1030, than below 1020. The question of an excess of urea is particularly important as bearing upon the metamorphosis of the tissues; and while I admit that my method of investigation cannot be regarded as conclusive, I hold the results to be sufficiently definite to discountenance the view of a frequent deficiency of urea in those cases.

With the greater facility for quantitative and qualitative analyses of the urine introduced by the volumetric method of Liebig and Neubauer, the determination of the constituents of the urine, and especially of the urea, will aid in the appreciation of the morbid phenomena.

But the great difficulty is that epileptics are always desultory patients, and being rarely confined to bed, it is more difficult to collect their secretions for the twenty-four hours, and thus fulfil all the requirements of science, than it is in many other diseases.

---

* See this author's interesting papers on Epilepsy in the *Med. Times and Gazette*, vol. xii., 1856, pp. 84, 133, 206, 279.

Though from the intermittent character of the disease we cannot expect ever to discover in the kidneys the causa proxima of the affection, the knowledge of any material and uniform deviation from the normal standard of so important a secretion, would necessarily aid in the comprehension of the changes induced by or associated with the paroxysm.

Scarlet fever is one of the diseases which is very apt to be followed by albiminuria, and it has also, with other eruptive fevers, the reputation of carrying epilepsy in its train. The inference would naturally be that the epilepsy was the result of the deranged state of the renal secretion. In the two cases of epilepsy which have fallen under my notice, and where the paroxysm followed so closely upon the scarlet fever as to justify our belief in a close relation between the two, no albumen was detected in the urine. As, fortunately, epilepsy rarely follows upon scarlet fever, and, as the above two cases show, its occurrence under such circumstances is not necessarily connected with disturbance of the renal function, we are compelled to look beyond this malady to find the predisposing cause of the epileptic seizure.

I have met with convulsive disease preceding and following measles; and I may mention that the severest case of opisthotonos that I have witnessed was attributable to suppressed scarlet fever, when after death an effusion of lymph was found upon the cervical portion of the spinal cord. In the analysis of cases I have, however, been careful only to include those that were indubitably epileptic.

Although the convulsions that usher in the eruptive fevers so frequently are not to be termed epilepsy, they are distinctly epileptiform; there is complete unconsciousness, violent struggling and evident arrest of the cephalic circulation; but they commonly pass off without leaving any trace, and may only be put down under the category of epilepsy if they persist after the disappearance of the eruptive fever. I shall take another opportunity of speaking of the doctrine of metastasis as applied to epilepsy.

The heart, the lungs, the liver, the spleen, manifest, as far as we know, no lesion which can in any way be brought into a causative relation with epilepsy. I have not detected any permanent derangement in the heart's action, even in prolonged cases of epilepsy; no such relation appears to exist between disease of the heart and epilepsy, as we know to prevail between chorea and heart-disease.

While I still hold to the general views expressed in the above passage, I may state that I have seen cases since the publication of the first edition which seem to bear out a remark of Dr. Todd's,* that many of the epileptic seizures which take place only or chiefly at night in elderly persons, have some intimate connexion with a diseased condition of heart, though I cannot adopt his qualification, "an altered condition of its muscles rather than of its valves." I should rather be disposed to submit for investigation, whether in these cases there is not a mere coincidence, the altered nutrition in the brain being the proximate cause, and that, again, depending upon a diseased condition of its blood-vessels.

An impaired state of the digestive powers is frequently associated with epilepsy, and the regulation of

---

* Clinical Lectures on Paralyis, Diseases of the Brain, and other Affections of the Nervous System, p. 307. London, 1854.

the diet of the epileptic is an important indication, which applies to all, but to none so forcibly as to children. But though a predisposing influence may occasionally be traced to continued insults being offered to the stomach, still this cause cannot rank higher than almost every other influence which carries with it a debilitating effect. Dyspeptics may be counted by thousands, while epilepsy can only number its victims by units.

More or less derangement of the bowels accompanies almost all varieties of epilepsy; costiveness, as I have already observed, habitually indicates the sluggishness of the intestinal movements, or a spastic condition of the muscular fibres of the tube, while the frequent presence of worms and the abnormal state of the fæces further demonstrate the unhealthy condition of the alvine tract. These derangements are so frequent, that I question the propriety of establishing, as Prichard has done, a distinct class of enteric epilepsy. Where the general health is much and extensively disordered, it is often a mere matter of individual interpretation whether we regard a given disease as more closely connected with the affection of one organ or another. I am, however, tempted to extract from Dr. Prichard's work a case which he gives under the head of enteric epilepsy, both because it illustrates the variety to which he applied the term, and because it is also an instance of a most remarkable case of cure of epilepsy in an idiot. The treatment is essentially the same in this case as that generally adopted by Prichard in enteric epilepsy, consisting in the use of oil of turpentine, which, according to him, exercises a peculiar sedative and tranquillizing effect upon the nervous system. The complication of epilepsy with any form of insanity is regarded by all physicians who have turned their attention to the subject as the most hopeless variety. Yet it would appear that a case like the following ought to encourage both the medical man and his patient never to despair, but to induce the former perseveringly to search out the secrets of nature:—

"Henry Parker, æt. 18, St. Peter's Hospital, January, 1818, an idiot, subject to epileptic attacks from infancy. He has sometimes three or four attacks in a day. At other times they occur once in a day; occasionally he escapes them in a week. He is insensible of their approach. He labours now under diarrhœa. His food passes off in an undigested state : his appetite is voracious; he goes about the house and yard picking up anything that can be eaten, and devouring it; he has been seen eating a cabbage-leaf from the ash-pit. He is emaciated, and his skin has a yellow tinge. He was sent to the medical ward. A warm bath twice a week and a nutritious diet were ordered for him. There is an eruption on his skin of the character of psoriasis. It occasionally subsides in the course of a night; the diarrhœa then becomes more troublesome, and when the eruption again appears the diarrhœa is relieved.

" Jan. 8. Ol. terebinth. ʒj. statim.

" Jan. 9. The dose purged him, and brought away undigested matter, but nothing like worms.

    Habeat ol. terebinth. ʒij. ter indies, et
    ℞ Pulv. rhei, hydr. c. cret. āā gr. v.
    Pulv. arom. gr. ij.
    post sing. dosibus tereb. sumend.

He took twelve powders; they relieved the stomach of the unpleasant sensation of heat produced by the turpentine. The turpentine has been given regularly, and he has been allowed full diet. The fits return now about once in seven days; they are not so violent as formerly, and last but a short time. The voracious state of his appetite continues. The yellowness of the skin is considerably diminished, and his general appearance much improved. The bowels are regular; the eruption disappeared gradually. The tongue is a little furred. He appears aware of the improved state of his health, and grateful to his attendants.

> ℞ Magnes. carb. ʒj.
> Pulv. rhei ʒij.
> Syr. papav. alb. ʒj.
> Aquæ puræ ʒvj.
> ʒij. ter die.

He took this mixture for some days, but it produced no improvement in the state of his appetite. The oil of turpentine was then resumed, and given twice in the day.

"June 13. Twenty-one days have now elapsed without any return of fits. He complains of headache. The pain is confined to the forehead. Leeches were ordered to be applied to the head about once in seven days, and the head to be kept shaved.

"September. The fits have gradually ceased. The paroxysms have been latterly much more slight; they lasted but a few minutes, and he was rendered aware of their approach by languor and headache. He is still in a state of idiotism.

"Jan. 1, 1819. He has experienced no return of the fits. He complains occasionally of headache, which is relieved by a purgative. He was discharged from the medical ward, and put upon the common diet of the house; till this time he had been allowed meat every day.

"Nov. 3, 1819. He continues in a much-improved state, though no medical care is taken of his health, and though he fares upon the common diet of the paupers in the house. He has had no returns of his complaint but once, when he had some very slight fits.

"Sept. 1, 1821. He has long been free from the symptoms of epilepsy."

This is not a solitary, though a rare, case. In Schroeder van der Kolk's valuable work* we find records encouraging us not to abandon hope, even in apparently hopeless epilepsy. He distinctly says that rare as is the recovery from epilepsy of long standing, with idiocy, a physician should not be too ready to despair, especially if the patient is still young.

The relation between the sexual organs and epilepsy is one that deserves our special attention, since it appears undoubted that the physiological evolution of these organs, no less than certain morbid states, are found so frequently associated with epilepsy as to justify the inference that they may stand in a closer relation than one of mere coincidence.

When speaking of the distribution of epilepsy over the different periods of life, I had occasion to observe its predominance about the time of puberty; it is essentially a disease of that period. The physiological character of puberty is the development of the sexual

---

* Professor Schroeder van der Kolk on the Minute Structure and Functions of the Spinal Cord and Medulla Oblongata. Translated from the original by W. D. Moore, A.B., M.B. New Syd. Soc. 1859, p. 269.

powers; anything that unduly promotes or interferes with that development is certain to give rise to violent reaction in the system. And considering the importance of the function and its manifold relations with the centres of innervation and circulation, it is in accordance with all our knowledge of physiology and pathology that it should be a frequent cause of reflex action. Hence, both in the male and in the female sex, an inquiry into the state of the sexual organs is an essential part in our investigations. Venereal indulgence is more frequently accused of being accessory to the epileptic paroxysm in the male; continence with a consequent undue excitement of the nervous system, is more commonly viewed as a cause in females.

But although the unanimous consent of all writers on epilepsy demonstrates the truth of the statement that in this disease the sexual organs are very frequently at fault, and indicates the necessity of our attention to their condition in order to judge correctly how we may apply our therapeutic proceedings, it is by no means determined in how far sexual derangements are to be regarded as a predisposing or exciting cause. An ancient medical proverb avers that "coitus brevis epilepsia est,"* we should therefore expect sexual intercourse to be a frequent cause of the disease. But although several writers record cases of epilepsy supervening upon the act of cohabitation, and I myself have seen such, the legitimate performance of the function

---

* The antiquity of this saying is proved by Galen remarking, in a discussion on the influence of cohabitation in the production of weakness and exhaustion, τίς γὰρ ἦν ἀνάγκη γράφειν Δημόκριτον μὲν εἰρηκέναι μικρὰν ἐπιληψίαν εἶναι τὴν συνουσίαν. The Democritus in question lived about 350 before Christ. Vide Galen Opera, ed. Küh. vol. xvii. a. p. 521.

is very rarely followed by such a result. This statement is corroborated by the fact, to which I can also bear witness from my expericnce, that marriage is at times followed by a cessation of the disease. Where the act of cohabitation is immediately followed by an epileptic paroxysm, it would be regarded as an exciting and not as a predisposing cause, since without the act the paroxysm would probably not have occurred. Our knowledge of the laws of reflex will explain such an event. The brother of a patient of mine was seized with his first epileptic fit the day after marriage, and died a few days later. An old gentleman, verging on seventy years of age, not otherwise subject to the disease, was seized with epilepsy, which was attributed by his wife to sexual intercourse. In the married state such events are painful, but give rise to no forensic complications; it is otherwise in cases where sexual intercourse induces epileptic convulsions and death in the unmarried, and where suspicion of violence or murder may arise.

Delasiauve[*] remarks that although he has made no special researches on the influence of sexual intercourse on epilepsy, he has noticed that some of his patients who continued free from attacks while in the hospital, again became subject to them as soon as they quitted it, and were then enabled to follow their sexual impulses. He refers especially to the case of one man who on four or five occasions, on his urgent solicitation, was allowed to return to his wife, and who came back each time, after a few weeks, with an exacerbation of

---

[*] Traité de l'Epilepsie, par le Docteur Delasiauve, Médecin des Aliénés Epileptiques et Idiots de l'Hôspice de Bicêtre, p. 1834. Paris, 1854.

his malady, and with symptoms of maniacal delirium, for which no other cause could be assigned but the exercise of his marital rights.*

The sense, however, in which sexual derangements are commonly found to induce epilepsy, is by enfeebling the system; by producing an excitability of the nervous system, an "impressionability," which, on the application of an exciting influence of sufficient strength, gives rise to the epileptic paroxysm. It is observed that in epileptics generally the sexual feelings are strong, so that in persons predisposed to the disease, the temptation to a vicious indulgence is probably peculiarly powerful. The inquiry into these circumstances is always fraught with considerable difficulty, as boys and young men are not easily induced to confess to masturbation, and unless carried to a great extent the vice cannot be positively predicated from their external appearance and demeanour. Young females who are guilty of the same fault can at all times only be suspected, as we possess no means of eliciting a confession. In either case the opinions of the parents go for very little, as they are too apt to be blind to such faults, or are misled by the positive assurances of their children, who at all times will rather confess their vices to a medical man than to a relative.

Females who have passed their teens at times complain of sexual irritation as manifested by leucorrhœal discharges, herpetic eruptions, and severe vaginal pruritus, which induces manipulation followed by epileptic

---

* For instances of the influence of sexual intercourse in the production of cerebral disease, though not of epilepsy, see Dr. Watson, "Lectures on the Principles and Practice of Physic." Fourth edition, vol. i. p. 497.

seizures; but these symptoms are often the results of diseased reflex action from the uterus and ovaries, to which the attention should be directed in the first instance.

In a person guilty of masturbation we generally notice a peculiar hang-dog expression; an unwillingness to meet the speaker eye to eye; a large sluggish pupil; a pale, livid hue and languid circulation of the surface, a general nervousness of demeanour: if we examine the urine, we often discover the patient to be labouring under azoturia, showing that an excessive metamorphosis of the tissues is perpetuated; varicocele, and a gleety discharge are frequently induced in men, and leucorrhœa may be sometimes attributed to this cause in females. Leuret found that in twelve out of 106 cases of epilepsy in the male, onanism was to be regarded as the cause of the disease, while Beau, in 242 cases of female epilepsy, traced onanism only three times.

Among the twenty-nine male epileptics of my second analysis of fifty-two cases, I find no less than nine in whom the sexual system was in a state of great excitement, owing to recent or former masturbation; a tenth should be added, who confessed to having very strong sexual passions, but asserted that he in no way at any time yielded to them; this, then, might be classed with those cases, certainly rare in man, in whom continence might be regarded as influencing the disease. It could not be said to have excited it, because in the particular case the fits had occurred in the first instance before puberty.

It is a common thing to be consulted by adult epileptics, who admit having formerly been guilty of self-abuse, but having abandoned the practice, continue

K

130 EPILEPSY AND EPILEPTIFORM SEIZURES.

to suffer from frequent seminal emissions, which are palpably in close relation to the epileptic seizure. These cases are very difficult to deal with. Where the masturbation is recent, and there is no serious hereditary or other complication, there is great hope that by inducing an abandonment of the habit a perfect cure may be achieved. As an illustration of the above remarks I adduce the following case:—

A. B., æt. 11, of florid complexion, but rather scrofulous appearance, and stated to be descended from scrofulous parents, the son of a surgeon in the country, was brought to me in May, 1859. He was a boy of impulsive emotional character, and had been subject to epileptic seizures, preceded by an aura passing from the middle finger of the left hand up the arm; the finger then became flexed, as did all the other fingers, and the arm itself was agitated and jerked from the side. The father had no doubt of his being perfectly unconscious during the seizures. The attacks themselves were preceded in the first instance, and subsequently, by an indisposition lasting one or two days, and marked by indigestion. The first attack was attributed to some dietetic excess at Christmas—a view which was strengthened by the passage (under the influence of purgative medicines) of considerable quantities of undigested food. Some irritation was also supposed to arise from the molar teeth, which he was cutting, and they were accordingly lanced. Worms were sought, but not found. On applying the galvanic current with a metallic brush to the left arm and hand, the sensation of an approaching paroxysm was at once produced, and he became pale; no further experiment was, therefore, made with that agent. As he bit his nails to the quick, and some abrasion had originally been noticed on the finger near the nail, from which the aura commenced, I recommended the tips of the fingers to be covered with a coating of gutta percha, and also a ligature to be worn on the fore-arm, so as to be able instantly to tighten it, and arrest the aura, an effect which was readily produced. In addition to these measures I advised the employment of tepid steel baths, and the internal administration of steel and cod oil. The attacks at the time of the first consultation occurred once in ten or twelve days, and were always preceded by a large deposit of lithates. The intervals now became lengthened out to six weeks. At the next visit, in October, I discovered that the boy was in the habit of perpetrating onanism. There was much costiveness, some incontinence of urine, and the latter generally presented a high specific gravity. He had given up the habit of biting his nails, and promised me solemnly, after a little private exhortation apart from his relatives, to abandon that of masturbation. I am not one of those who disbelieve in the power of medicines in influencing epileptics; in fact, the present case affords strong evidence in favour of that view; the fits had increased under the previous treatment with iodide of potassium, but diminished under the tonic plan. Now, however, they ceased altogether, and have not returned since. At least, I infer this; for a letter from the father, dated November 15th, 1860, states that there had been no return of the fits since the 28th September, 1859, or nearly fourteen months; and I am well assured that I should have heard again had they recurred since.

In illustration of the statement regarding the influence of seminal emissions in the adult I may quote the following abridged case from my memoranda:—

C. D., a gentleman, æt. 23, of delicate habit, and presenting a scrofulous taint, has been subject to epilepsy for about two years, preceded by an aura, beginning in the little finger of the right arm, and mounting up to the head; after which general spasms of all the extremities, with unconsciousness, follow. The attacks generally occur in the evening or in the morning, and there have been about fifteen severe fits altogether; small attacks of convulsive action of the extremities, without unconsciousness, not being reckoned. He attributes his attacks to masturbation, which he used to commit, though he has abandoned it completely. But he continues to be subject to very frequent involuntary emissions, which are always followed by a seizure. He believes that recently he has frequently had attacks in his sleep, from finding that there had been a discharge in the morning, and from his then experiencing all that feeling of oppression and exhaustion which characterizes the sequelæ of the acknowledged epileptic fit. Besides that, he has found his tongue bitten on several occasions in the morning on waking up. The remaining details of the case have no special bearing upon the questions under consideration.

In females, the relation of sexual function is shown in a more marked manner by the frequent derangement in the menstrual orgasm on the one hand, and by the concurrence of the epileptic seizures with the period of menstruation on the other. Epilepsy in the female at and after puberty is very frequently accompanied by some derangement of the catamenia: either they are altogether arrested, or they are irregular and scanty, or they are (but less frequently) menorrhagic; or, again, there is constant and profuse leucorrhœal discharge. But even where no marked abnormity is to be detected in the character of the menstrual flow, we still, in women, often find a definite periodic character imparted to the epileptic seizures, which precede or follow the catamenia.

In the first edition I laid some stress upon this relation, but my enlarged experience induces me to dwell even more urgently upon it than I then did. My reasons for this will be apparent to the reader on perusing the subsequent statements on the subject, and if he would be at the trouble to compare these passages with what I stated at page 110 in the first edition, he would see that I was not without justification.

It is not, however, to be concluded that the physician is therefore to direct his entire attention to the sexual function. In the female this at all times more or less governs the whole system, and epilepsy does not escape the general law; hence, while in female affections of every kind it is never to be lost sight of, its condition in epilepsy may not be regarded in any other light than as indicative of a general derangement of the system, coincident with the epileptic paroxysm, and though not causing, yet favouring, its occurrence.

In most other affections of the female, the restoration of the menstrual period to its normal condition, or its first appearance at puberty, tends to diminish or cure other coexistent morbid conditions; in epilepsy this relation does not appear to prevail to the same extent. Herpin* states that he has not met with a single instance in which the appearance of the catamenia has induced even an improvement, not to say a cure, of the complaint. This very unfavourable experience does

---

* Du Pronostic et du Traitement Curatif de l'Epilepsie. Par Th. Herpin. Paris, 1852.

not coincide with my own and that of some other observers; the *laissez-aller* practice which would result from the adoption of such views would be opposed to all sound pathology. They would tend to induce medical men to neglect the first indication in all affections of the female—viz., of restoring the catamenia to their normal condition. Dr. Prichard* gives a long chapter on what he terms uterine epilepsy; a form which he describes as mainly affecting young females of a sanguine temperament, ruddy complexion, and light hair, and resulting from a suppression or some other derangement of the catamenia. He details fifteen cases, and refers to four others which have occurred in his practice. One of the cases proved fatal, and after death the left lateral sinus was found blocked up through its whole extent by an old deposition of lymph which had become organized. Almost all the others were cured by repeated venesection, warm-baths, and the internal administration of oleum terebinthinæ. The results obtained by Dr. Prichard are clearly at variance with the categorical affirmation of Herpin. Still the immediate bearing of the state of the uterine function upon epilepsy in a large number of cases is imperceptible.

On examining my records with reference to this point, I find that the cases of females subject to epilepsy may be divided into three classes; 1. Those who have not attained puberty, and in whom, therefore, the sexual organs may be supposed not to exert any material influence on the economy at large; 2. Those in whom the sexual organs have arrived at maturity; 3. Those who have passed the climacteric, and who,

---

* A Treatise on Diseases of the Nervous System, part i. p 148. London, 1822.

therefore, in regard to the sexual organs, present an analogy to the first class. If we add the twenty-seven cases recorded in the first edition to the twenty-three yielded by the recent analysis of fifty-two cases of epilepsy, we find the total of fifty female cases arranged under the three categories in the following proportions:—

Before appearance of the catamenia.	During the continuance of the catamenia.	After the cessation of the catamenia.
10	37	3

Under the first class I find one in whom the appearance of the catamenia is reported to have caused an arrest of the fits.

The second category is the one that most concerns us now, and on analysing the thirty-seven cases, I find that in thirteen the catamenia are reported to have been normal, in seventeen there is definite evidence of some derangement of the catamenia or other sexual disorder, and in seven there is no specific statement on the point. However, it is reasonable to add that among those reported to have been regular, there are three in which it appeared as if there was some relation between the sexual function and the fits; because one stated, that though formerly, not regularly menstruated and of late regularly, the fits had only supervened since the catamenia had been normal; another averred that the fits were aggravated by the approach of the period; and a third that, although regular, the seizures occurred monthly before the catamenia. In the same way, in the seven cases which give no definite report as to the catamenia, we are not left without some information as to the influence of the sexual organs on the epilepsy. Six out of the seven were married women, one of whom stated that marriage had diminished the attacks, another became epileptic during pregnancy, and a third ceased to be epileptic during pregnancy, but was worse again while suckling her child.

In the case of the seventeen patients who complained of some definite derangement, it may be useful to enumerate the symptoms as they were stated. 1, æt. 17, amenorrhœa, only once slightly menstruated. 2, æt. 15, once slightly menstruated. 3, æt. 17, catamenia irregular and scanty. 4, æt. 34, catamenia regular with leucorrhœal discharge. 5, æt. 42, catamenia somewhat irregular. 6, æt. 32, catamenia regular but scanty; the fits occur chiefly immediately before or after the catamenia. 7, æt. 16, catamenia had not appeared at sixteen years of age. 8, æt. 24, catamenia scanty, the fits worst on their cessation. 9, æt. 41, intense persistent pruritus vaginæ. 10, æt. 20, catamenia often scanty and delayed; leucorrhœa, fits often at period. 11, æt. 18, catamenia irregular, attacks commonly at period. 12, æt. 46, catamenia irregular and scanty, fits chiefly at catamenia. 13, æt. 19, catamenia irregular and scanty. 14, æt. 17, some uterine derangement. 15, æt. 16, catamenia irregular, fits said to be worse at catamenia. 16, æt. 20, first fit occurred after a sudden arrest of the catamenia. 17, æt. 32, catamenia menorrhagic, frequent leucorrhœa, remarkable periodicity of fits in relation to the catamenia.

It is impossible to see any number of epileptic females and not to recognise the existence of some relation between their disease and the sexual organs. It is more difficult to determine the exact nature of the relation; whether it is simply indicative of a common debilitating cause, or whether it is to be regarded as the direct excitant. When we find the sexual derangement, it is our duty to combat it and remove any anomaly if possible. We may, by attending to this indication, cure the disease, or at all events alleviate and diminish it; but, unfortunately, in many instances we shall be foiled in our endeavours, and though the uterine malady is cured the epilepsy persists. I shall return to this subject in the chapters devoted to treatment, but I will quote one case in which there was well-marked periodicity, and where there is ground for believing that a cure, or at least material improvement, would have resulted from a more conscientious pursuance of the necessary measures on the part of the patient. I may mention, *en passant*, that it was one of the cases in which the bromide of potassium failed, although apparently well suited for it.

E. F., a married lady, æt. 32, the mother of several children, had been subject to epilepsy for twelve years, and for four or five years past they had occurred with great regularity; a fit occurred almost always at or about the catamenia, and when at other times, it was first at the middle of the intervening period. The attacks took place indifferently at all times of the day, generally when the patient was in a state of inanition. She had no premonitory symptoms; screamed and became totally insensible, in which state the body was violently convulsed. The catamenia were menorrhagic, there was frequent leucorrhœa, much bearing-down, difficult micturition, and occasional pruritus. The cervix uteri was soft and flabby, the os large and patulous, with numerous cicatrices, the vagina large and leucorrhœal. The rest which was enjoined as one of the most essential points of the treatment, was not maintained; and in spite of all warning, the patient, a lady of very impulsive character, frequently subjected herself to excitement and fatigue, which was too manifestly connected with the subsequent attack to leave a doubt that they might to a great extent have been warded off by proper care.

In speaking of the sexual derangements of women in connexion with epilepsy, it is well to remember, that in every chronic disorder of the female, the uterus and ovaries are apt to present some abnormal condition; but the same cannot be said of the male sex, hence the numbers indicating a relation between epilepsy and sexual anomalies in the latter are proportionately of greater importance.

In connexion with the present subject, it will be suitable to say a few words on the influence which marriage exerts upon epilepsy. That it prevents masturbation in men, and tends to relieve many disorders of the sexual functions in females, is an undoubted fact; but, as we have seen that the marital act itself may become an exciting cause of epilepsy, and as we know that the hereditary influence of the disease is great, we ought, as a rule, to discountenance the marriages of epileptics, as well on account of their partners as on account of their offspring, unless the long time that has elapsed since the occurrence of a paroxysm justifies a hope that the morbid taint is quiescent, if not extinct.* Herpin, who has taken much pains to determine the inquiry by facts, has raked up† three

---

* Prichard takes a different view on this point; see his "Remarks on Uterine Epilepsy," l. c., p. 217, where he says: "Pregnancy generally removes disorders connected with disorders of catamenia; but even if pregnancy should not take place, I am very much disposed to believe that diseases of this kind would generally be removed by marriage."

† Du Pronostic et du Traitement de l'Epilepsie, p. 520. 1852.

doubtful cases from unknown authors,* which have been cited to prove the beneficial influence of marriage in this disease. It is manifest that the difficulty of meeting with instances which establish the point, sufficiently demonstrates the truth of the general law that marriage is not curative in epilepsy. The fact that the greater number of epileptics are *célibataires*, is due to the circumstance that the affection makes its first appearance most frequently at puberty.

Hébréard in 1813 found that of 162 epileptics at Bicêtre,—

119 or 73·5 per cent. were bachelors,
36 „ 22·2 „ „ married,
7 „ 4·3 „ „ widowers or divorced.

Moreau found, in 1821-22, that, of 240 female epileptics,—

142 or 59·2 per cent. were spinsters,
32 „ 13·3 „ „ married women,
17 „ 7·0 „ „ widows,
49 „ 20·4 „ whose condition was unknown.

But, as Moreau himself observes, these numbers prove nothing as to the question at issue, " because the great majority of these patients do not marry."

When our advice is asked as to the propriety of an epileptic patient marrying, we must take into consideration the entire aspects of the case before we come to a decision which so materially affects the future of two persons at least. The severity of the disease in the

---

* Hoffmann gives a case of the kind, Opera, tome iii. p. 21. I may point to one of my own cases, in which the patient, a widow of thirty-eight, stated that she had had fits from time immemorial, but had been freer from them since marriage and childbirth than formerly.

individual, the amount of hereditary taint, the influence of treatment, the probability of an arrest of the complaint, the health of the person whom our patient is to marry, and the question of consanguinity, are all matters that must be earnestly weighed in our mental balance. One patient who came under my care on account of crural neuralgia at the age of twenty-three, had from the age of fourteen to twenty-one been subject to severe epileptic fits, in which she fell into the fire and bit her tongue, and for which she was under eminent physicians to no purpose. She married at twenty-one and the fits ceased. The mother of an epileptic lad, for whom I was consulted, was herself epileptic before marriage and continued so till the birth of a daughter, when the fits ceased and continued in abeyance for fifteen years, up to the time of my seeing her. It is true, the disease was transferred to the children in this case, and may so far be an argument against the marriage of epileptics. Where the parties about to contract marriage are nearly related, medical men should always refuse their sanction; but when, as in a case now before me, in addition to such consanguinity, there is epilepsy on both sides, I hold that the Legislature ought to intervene to prevent the misery which must follow.

From the preceding remarks it will be inferred that I would only put a qualified veto upon the marriage of epileptics. I may further illustrate my views on the subject by the abridged details of the following case, together with the certificate which I considered myself justified to give in reference to it.

A gentleman from a western county, a florid, robust-looking person, aged twenty-one, had for fifteen years been subject to epileptic attacks, when he consulted me in

July, 1859; they occurred on an average once in three or four weeks, they came on without premonitory symptoms, and he had only once bit his tongue. The fits occurred only by day, lasted ten or twelve minutes, and were followed by sleep, from which he awoke completely refreshed. The attacks took place irregularly during the day time, but never after 11 p.m.; excepting the fits, the general health was good, the pulse quiet and slow, and no visceral lesion. In consequence of the treatment, the intervals were gradually lengthened out to three and seven months, and an opportunity for marriage presenting itself, I was asked to give my opinion in writing in June, 1860; the medical attendant stated at the time that he could report favourably. "I can safely say that his appearance generally is so much improved, that you would scarcely know him again. With regard to the epileptic attacks, he has had only one in seven months, and that a few days since, of a very slight character; it happened after a day's rook-shooting, followed by a supper at his own house. I did not see him in it, but am told that he was insensible only for a very short time, and that he returned to the supper-room half an hour after perfectly collected, and remained the rest of the evening with his friends."

The following is a copy of the certificate I gave, dated June 5, 1860:—

"I am of opinion that as the epileptic seizures to which Mr. L. M. has been subject during the last fifteen years, have decreased very much in frequency since I last saw him in July, 1859, there is great probability of the attacks being entirely arrested. Formerly they occurred once or more times in the month; since consulting me in 1859, he has only twice been seized, and each time with less severity. Each time, too, the attacks were brought on by definite causes which ought to have been avoided, and which Mr. L. M. knows how to avoid in future. Although it is impossible to pledge myself to a complete cure, I have no hesitation in saying that, considering all the circumstances of the case, Mr. L. M.'s prospects are extremely favourable. I cannot but think that the perfect regularity of domestic life would be further conducive to securing Mr. L. M.'s health, and therefore, on medical grounds, I cannot dissuade him from marriage, but would rather recommend it. While saying so much, however, I assume Mr. L. M. will continue to attend to those general rules of diet and regimen which I have laid down for him."

Although I do not doubt that continence *may* by itself give rise to epilepsy, it is excessively difficult to obtain reliable information on the point, and the cases I should imagine to be rare, at least in men. The older authors dwell more upon this cause than recent writers. Those curious upon the point will find an assemblage of cases showing the effects of continence, among which epilepsy is included, in the first volume of Tissot's works;* partly quoted from Hoffman, Zacutus, and others, partly supplied from his own experience.

In women I should be disposed to admit the more frequent operation of this causative influence than in men, both from the greater morality of the former and from the greater susceptibility of their whole system to all impressions conveyed through the organs of reproduction.

---

* Œuvres de Mons. Tissot. Tome premier, contenant l'Onanisme, p. 224, seqq. Lausanne, 1790.

CHAPTER VI.

The exciting causes of epilepsy, primary and secondary — Psychical, physical influences — Diurnal changes — Dr. Smith's experiments — Syphilis — Sleep — Intemperance — Centric and eccentric epilepsy — Dr. Short's, Dr. Darwin's, and La Motte's cases.

The influences which we have hitherto considered as bearing some definite relation to epilepsy are commonly enumerated among the predisposing causes of the affection. Another class of so-called causes are entitled exciting, because they are supposed to exercise a more direct influence upon the production of the complaint. From their more immediately preceding the paroxysm there is better ground for connecting them with the latter than often applies to the predisposing causes, which are more vague, while their interpretation is often more fanciful. In a large number of instances neither the patients nor the physician can detect any peculiar circumstance to which the outbreak of the paroxysm can be immediately attributed. It seems as though the electric battery became overcharged,* and the discharge was effected in the form of the epileptic seizure

---

* According to Schroeder van der Kolk, this is an actual account of what takes place, inasmuch as he holds that the ganglionic cells of the medulla oblongata really act upon the same principle as electric batteries. See Professor Schroeder van der Kolk on the Minute Structure and Functions of the Spinal Cord and Medulla Oblongata, &c. New Syd. Soc. ed., 1859.

simply to relieve the excessive tension. If the comparison is just, it is intelligible that no palpable occurrence is required to elicit the current, but that the mere approach of an opposite pole may suffice for the purpose. And this would really seem often to be the case; for in spite of the most careful search, we are unable, in a large number of instances, to trace anything approaching to the character of an exciting cause in the individual paroxysms. In fact, in many cases we observe not only no exciting cause, but we see our patient, in better health than usual, suddenly prostrated by this dire disease, as if by a thunderbolt shot 'from a clear blue sky.

We must distinguish, too, between what may have been the apparent exciting cause when the disease first made its appearance, and those exciting causes to which subsequent attacks may have been individually due. They are by no means identical. Thus, fright is a frequent cause of the first seizure, but may never operate again; but the impulse having once been given, future attacks may follow by the law of periodicity, without any specific excitant; or they may be due to an indigestible article of diet, a fatiguing walk, or some emotional influence. Or, again, the first attack may be reasonably attributed to the presence of tænia, the parasite may be removed, and the tendency to the epilepsy still continue subject to any morbific influence to which the individual is exposed.

The mode in which the exciting cause operates I should regard as twofold: either it acts by direct stimulation of that part of the nervous centres which is immediately concerned in the production of the paroxysm, or it operates by indirect stimulation and irritation of the same part according to the laws of reflex.

PSYCHICAL AND PHYSICAL CAUSES. 145

Psychical causes would for the most part come under the former category; while the physical causes would come under both, some acting by direct stimulation, some by reflex. The influence of sleep in producing epilepsy, if acting by the contact of highly carbonized blood on the mesocephale, would be a direct stimulus, while if it acted by inducing sexual orgasm it would be by reflex; the interpretation of course depending upon the correct appreciation of the observer.

I find that an analysis of 104 cases of epilepsy observed by myself shows a definite exciting cause to have been traceable at the outbreak of the disease in fifty-six; among these, fifteen were of a more or less purely psychical, forty-one of a physical, character. They are as follows:—

Psychical.	
Fright	6 cases.
Mental work	5 ,,
Anxiety	4 ,,
	15

Physical.	
Otorrhœa	2 cases.
Scarlet fever and fever	4 ,,
Dentition	3 ,,
Operation	1 ,,
Gastric and intestinal derangement	5 ,,
Injury to head	2 ,,
Pregnancy	1 ,,
Lead poisoning	1 ,,
Evolution of puberty	3 ,,
Scorbutic condition of blood	1 ,,
Uterine derangement	9 ,,
Masturbation and venereal excesses	6 ,,
Intracranial disease	1 ,,
Psoriasis	1 ,,
Spinal rheumatism	1 ,,
	41

It may be interesting to compare with my own summary of causes those elicited by other observers. I extract from Moreau's Memoir the following two tables, in which the influence of psychical affections in the production of epilepsy is particularly apparent.

The first table, by Leuret, comprises 106 male epileptics, in whom the exciting causes traced were—

Fear	35 times.
Anger	2 ,,
Drunkenness	6 ,,
Falls	3 ,,
Onanism	12 ,,
Poverty	2 ,,
Psoriasis	1 ,,
Insolation	1 ,,
Chill	1 ,,
Unknown	39 ,,

The second table, drawn up by M. Calmeil, includes 240 female epileptics:—

Psychical Causes.

Fright	51 times.
Fear	31 ,,
Painful impression	15 ,,
Disappointment	11 ,,
Sight of an epileptic	2 ,,
Rape	9 ,,
Unkindness	5 ,,
Anger	4 ,,
Joy	2 ,,
Grief	12 ,,
	142

Physical Causes.

Mercury	1 time.
Childbirth	2 ,,
Suppressed epistaxis	2 ,,
Suppressed catamenia	1 ,,
Causes that were not appreciated	21 ,,
Critical age	2 ,,
Poisoning by camphor	1 ,,
Severe operation	1 ,,
Onanism	3 ,,
Without cause	64 ,,
	98
Psychical	142
	240

Among the exciting causes I may mention one which is not included in any of the tables; it operated in the case of a governess who was under my care for amenorrhœa and symptoms connected with uterine disturbance. One morning during a severe thunderstorm she was seized with a series of severe epileptic fits; she was completely unconscious, was convulsed, foamed at the mouth, and bit her tongue and lips. She attributed the seizure to the influence of the thunder and lightning, because two years ago she had a similar attack during a thunderstorm, and because, although not frightened, she experiences peculiar sensations at the time.

Whatever value we may attach to analyses of this kind, one indubitable fact results—viz., that psychical causes operate much more frequently in the production of epilepsy among females than among males; the proportion in the first table being 34·9 per cent., in the second 59·1 per cent.; a conclusion which entirely coincides with the physiologically greater "impressionability" of females. Nor can it escape attention that the psychical affections are, with trivial exceptions, those that would depress the nervous and vascular energy—a point which must be borne in mind in directing the treatment of epileptics, as it is quite as much the duty of the physician to operate on the body through the mind, as upon the mind through the body. It would be more easy to raise objections to most of the other exciting causes mentioned, since their importance is rather a question of individual opinion. Nor must it be overlooked that, in my own cases as well as in those quoted, there are a large number in which no cause could be discovered, although both patient and medical man are naturally anxious to trace some tangible circumstance, which may be supposed to have been instrumental in the production of so grievous a malady. Numerous as are the causes adverted to by the French authors, it is surprising that two are not mentioned which often exercise a powerful influence in the production of epilepsy—syphilis and dentition. It may at times be difficult to determine whether to class them with predisposing or exciting causes ; and another reason why dentition is often overlooked in the causation of epilepsy is the fact that many medical men regard the convulsive fits occurring during the period of dentition as essentially distinct from epilepsy. I am of opinion that both the relation of infantile fits to the epilepsy of later life, and the characters of the fits themselves, are a proof of the pathological identity of the two affections. I have not, however, in deference to prevailing opinion, in my tables mixed the two together, and therefore reserve some remarks on the subject for a later period.

With regard to syphilis, it is also difficult to say whether it acts as a predisposing or an exciting cause ; under different circumstances it probably acts differently ; thus it may merely enfeeble the body by depraving the blood, and thus lay it open to the influence of some other exciting cause, or it may induce epilepsy by giving rise to the formation of syphilitic disease of the cranial bones and direct injury to the brain. But here, as elsewhere, it is probable that some peculiar state of the nervous system must predispose the individual. Thus C. B., a man aged thirty-two, who had been syphilitic six years before he came under treatment, since which time he had not enjoyed the good health he had previously, suffered from a series of abscesses, and a year before presenting himself for advice became epileptic, the fits recurring about once a month. The fact of his wife having had many miscarriages confirmed the opinion that the syphilis had thoroughly undermined the constitution, although, at the time of consultation, no positive proof could be adduced of the poison still existing in the system. The patient stated that he always had perfect health before the syphilitic infection, so that it was fair to regard the syphilis as closely connected with the epileptic seizure. Still, on close questioning, it was ascertained that he had had one fit when he was fifteen years of age, which was sufficient to prove that there was some other taint or weakness in the system capable of inducing epilepsy.

It would not be right to conclude the consideration of the exciting causes of epilepsy without once more adverting to the influence of sleep in its production. This is a point to which patients are not likely to advert spontaneously as a cause of the disease, because it is not sufficiently tangible; but to medical observers it is so evidently an adequate and efficient cause in many instances, that I am surprised that among the many classifications proposed, it has not been made the basis of any one, nocturnal and diurnal epilepsy being manifestly distinct in their rationale. Physiologists are unable to tell us the actual changes that take place in the cerebral circulation during sleep; they differ as to whether the brain is in a more or less congested state than in the waking condition;* but that the whole pro-

---

* The most recent experiments by Mr. Durham (Guy's Hospital Reports, third series, vol. vi. p. 149) seem to indicate that in healthy sleep there is no venous pressure, at the same time they show that any interference with the return of the venous blood from the brain is at once followed by coma and convulsions. One of this author's conclusions, besides being corrobora-

cess of nutrition and metamorphosis is different during the sleeping and waking state is demonstrated.

One thing is certain, that the respiratory process is carried on with less vigour, and that the blood reaches the brain in a lower state of oxygenation. If we adopt the conclusions arrived at by Dr. Smith, we must admit that at the same time the pulse is reduced;* we then see in the influence of sleep two elements introduced which directly affect the nutrition of the brain—viz., the blood becomes more charged with carbonic acid, and the heart acting with diminished contractile power, the blood circulates with less rapidity through the brain. The former circumstance is manifestly the main difference between the waking and sleeping state, while the influence of the latter upon the brain would be promoted by the recumbent position, as it would tend to increase the tax upon an enfeebled heart, and thus induce congestion and stasis. Dr. Smith infers from his numerous experiments, that "vital actions are at their maximum in the day, and their minimum in the night, and the action of the one and the negation of the other may be in excess or defect, and need regulation. Thus the action of the day may be in excess from great sunlight, and from too much or too fre-

---

tive of my views, has so direct a bearing upon the argument used above, that I quote it verbatim:

"The more the subject is considered, the more evident will it appear that interference with the return of blood from the cranium causes torpor, not so much by pressure of distended veins upon the brain, as by the hindrance which necessarily arises to the due supply of arterial blood, and by the presence in the vessels and tissue of an accumulated proportion of carbonic acid." (*Loc. cit.* p. 156.)

* See the admirable researches of Dr. Edward Smith on the Hourly Pulsation and Respiration in Health, Med.-Chir. Trans. vol. xxxix. 122-6.

quently repeated food, or too variable by long intervals between meals; whilst the negation of the night would be too great if the food taken in the day had been in defect, or the daylight had been absent. So the former may be in defect from the absence of light, food, and wakefulness, and the latter in excess, from the presence of those influences." These considerations are worthy of special examination in reference to epilepsy, and whether or not the particular facts and conclusions of Dr. Smith be made the basis of our theory, there can be no doubt of the decided influence of diurnal variations upon the epileptic paroxysm through an effect exercised upon respiration, circulation, and change of tissue, apart from pathological conditions.

I have alluded to position as influencing the cerebral circulation during sleep. I cannot do better than submit to the reader the following extract from Dr. Salter's* valuable work on Asthma in illustration of this point. He certainly is seeking only for an explanation of the mode in which asthma may be induced by sleep; but the connexion of the condition to which he adverts with the state of the cerebral circulation, is too obvious to require special comment. Moreover, nothing more strongly illustrates this relation than the fact, that epilepsy and asthma are interchangeable diseases, or, to employ an ancient term, that by a process of metastasis† one may be substituted for the other in the same individual.

---

* On Asthma, its Pathology and Treatment. By Henry Hyde Salter, M.D., F.R.S., Fellow of the Royal College of Physicians, &c. London, 1860, p. 64.

† Salter, *loc. cit.* p. 44, gives a remarkable case of the kind, to which, as to the whole doctrine of metastasis in relation to epilepsy, I shall take another opportunity of alluding more fully,

Speaking of the reasons of asthma coming on at night-time Mr. Salter remarks, that they are twofold: "One is the horizontal position of the body, the other, the greater facility with which sources of irritation, and, indeed, any cause of reflex action, operate during sleep than during the hours of wakefulness. The first cause acts thus:—When a person lies down, and goes to sleep, the recumbent position favours the afflux of blood to the right side of the heart, and therefore to the lungs; in addition to this, the position of the body places the muscles of respiration at a disadvantage, especially the diaphragm, against the under surface of which the recumbent position brings the contents of the abdomen to bear; to this may be added the diminished rate at which the vital changes go on during sleep; and lastly, the lowered sensibility of sleep, which prevents the arrears into which the respiration may be getting from being at once appreciated. Here I think we have a sufficient explanation both of the time at which the attack generally comes on, and of the amount of dyspnœa that may accumulate before the asthmatic is roused from his slumbers. He goes to bed perhaps quite well; the position of his body and the torpor of sleep soon throw his lungs into arrears, and they become congested; this goes on for some time gradually increasing, without producing any particular effect; but, by and by, this pulmonary congestion reaches such a pitch that it becomes itself a great source of local irritation, and gives rise to asthmatic spasm. This, in its turn, cuts off the supply of air, and increases the congestion, and the asthma and the congestion—the cause and the effect—mutually augment one another, till they produce such an amount of dyspnœa as is incompatible with sleep, and the patient wakes with all the distress of an asthmatic paroxysm full upon him."

Although the physiology of sleep enables us to form a reasonable explanation of the manner in which nocturnal epilepsy is produced, it must be remembered that it is applicable only to a limited number of cases, and only so far affords a clue to the general theory of the production of epilepsy that it suggests the very close connexion between the state of the circulation or of oxygenation, and the paroxysm.

I would infer from the preceding remarks, and from the result of every-day observations on children and adults when asleep, that the state of the brain during sleep is one, if not of actual congestion, still one in which this condition is more readily induced than in the waking state; and that when the nerve power is depressed, and new causes of congestion arise, those changes of nutrition result which induce the epileptic paroxysm in an analogous manner to that which, *mutatis mutandis*, excites the asthmatic attack.*

Into the general question of the variation in the quantity of blood within the cranium I do not think it necessary to enter. Dr. Burrows has completely † set

---

* In further support of the views above expounded, the reader is referred to the article "Sleep," by Dr. Carpenter, in Todd's Cyclopædia of Anatomy and Physiology.

† On Disorders of the Cerebral Circulation, &c. By George Burrows, M.D. London, 1846. I may state that I have repeated Dr. Burrows' experiments on rabbits, and that the results, as shown by drawings made at the time, are identical with his own. See also on this question the experiments of Donders, and of Kussmaul and Tenner; the latter (*loc. cit.*) distinctly state that the quantity of blood in the cranial cavity can, by way of experiment on the living subject, be considerably increased or diminished. Donders (Virchow's "Handbuch der Pathologie," bd. i. p. 111. and New Syd. Soc. vol. v. p. 39) having trephined the skull of a rabbit, and closed the opening air-tight with glass,

this matter at rest, which would never have excited so much controversy but for the misunderstood statements and experiments of Dr. Kellie. The concurrent testimony of all pathologists, daily observation in health and disease, and experiments upon animals, prove that such variation does take place; while those who have not the opportunity of making investigations for themselves can scarcely fail to be convinced by the clear exposition of facts and the calm logical reasoning of Dr. Burrows.

Some persons, otherwise in perfect health, suffer from spasmodic movements at the moment of going to sleep; in all we are able to induce automatic movements during sleep; and there is good reason for believing that persons in whom automatic movements are spontaneously induced during sleep—that is, who show more or less tendency to somnambulism, in which, while sensation is in abeyance, the automatic movements are continued—have a tendency to epilepsy. Sleep-walking, however, or, as Sir Henry Holland defines somnambulism, "a dream put into action," is a common occurrence in young children, and is by no means a necessary indication of approaching epilepsy. It is merely to be regarded as a proof of an excitable nervous system, and may be taken as a warning to parents and guardians to protect such children more carefully than others from all undue physical and mental excitement.

Some authorities hold that there is always an increase in the amount of blood in the brain during

---

demonstrated a change in the diameter of the vessels of the membranes as well as of the brain itself under varying influences.

sleep, and it has been suggested that the choroid plexuses become turgid. When congestion takes place in the brain in sleep, it must be from impairment of the respiratory process, and the blood must then become unusually dark or venous. This is the view held also by Sir Henry Holland,* who, in his admirable chapter on sleep, observes: " There is reason to suppose that such effects depend on the proportion of venous blood present in the cerebral circulation, either from congestion in the great veins or from imperfect arterialization in the lungs."

Among the exciting causes of epilepsy it is necessary to allude especially to intemperance. My own experience does not enable me to say much on the subject, although I have repeatedly seen cases in which the employment of fermented beverages was evidently of disadvantage to the patients, inducing and promoting the attacks. In the army this is alleged to be a common excitant of the paroxysm, and may be supposed to act by increasing the venosity of the blood, and thus inducing congestion of the brain. Dr. Leuret† found that six out of 106 epileptics attributed their seizures to intemperance.

It is a question whether the circulation in the spinal cord at night does not also deserve especial consideration in regard to the production of spasmodic affections. It can scarcely be doubted that the seminal discharges which occur during night depend in part upon a physical influence exerted upon the spinal cord, as position and

---

* Medical Notes and Reflections. Second edition, p. 452, 1840.

† Recherches sur l'Epilepsie (Archives Générales de Médecine, iv. série, tom. ii. p. 2).

the texture of the bed are known to determine their occurrence. The relation between sexual excitement and epilepsy has already been adverted to; but we must distinguish between the consequences of the normal performance of sexual intercourse, which involve the brain as much as the spinal cord, and the abnormal excitation occurring in sleep from the possible effects of spinal congestion. The nocturnal enuresis of young children belongs to a similar category as the seminal discharges of the adult; and in them also the tendency to convulsive affections predominates in sleep.

If the view is correct that the greater venosity of the blood is closely connected with the production of nocturnal spasm, how urgent a reason, in addition to all the other arguments that may be adduced, for attending to the maintenance of adequate ventilation in our sleeping apartments.

In reconsidering the causation of epilepsy, and briefly recapitulating the preceding arguments, we would insist upon the danger of too minute classification. There is always a risk, in a science like medicine, in attaching too much value to a term: this applies in the present instance to the relative value of predisposing and exciting causes, which we may rigidly distinguish upon paper, but which cannot always be as clearly kept asunder in the observation of a diseased individual. We have seen that, in many instances, it is impossible to trace a definite influence of either kind. But as far as the evidence goes that has been collected, there appears to be no room for doubt that most of the causes that are productive of epilepsy, operate by enfeebling the system at large; and, by impoverishing the blood, lay open the nervous system more particularly to injurious impressions, which in health would leave no effect.

The relative importance of the predisposing and exciting causes, and the reality of the distinction, are shown most emphatically in those forms of epilepsy which have been termed eccentric, because on the removal of the exciting cause the disease has ceased. Not one of the circumstances which in some individuals are productive of epilepsy are necessarily or generally so, but are met with very commonly in other individuals without ever giving rise to any form of spasmodic disease. Thus no disorder of the primæ viæ is more common in children than that which gives rise to parasitic animals; but the frequency of epileptic convulsions due to this cause is by no means proportionate to its frequency: and yet it would be careless in any case of epilepsy not to inquire into the point, because the removal of the worms, whether ascarides, lumbrici, or tænia, is demanded under all circumstances for the restoration of health, and numerous cases are met with in which epilepsy is arrested in consequence.

If we fail to discover an exciting cause, it is taught that we have to do with a diseased condition immediately affecting the brain, and the disease is then called centric epilepsy. But we shall see that here equally there is scarcely a morbid condition which has not been found in connexion with epilepsy, while every one of those pathological states occurs much more frequently independently of epilepsy. It would therefore appear more in consonance with observed facts to regard epilepsy as an affection invariably dependent upon some hitherto unexplained derangement in the nervous system, often dormant for years, and even for life, unless the exciting cause comes into operation.

If this view be correct, a distinction between essential and non-essential epilepsy cannot be said to exist; but wherever a paroxysm has occurred, we should assume the same peculiarity of the nervous system to prevail in a stronger or feebler degree, and the difference would be mainly in the agent which roused its susceptibility into action. To revert to a former simile, the diathesis may be compared to combustible material of greater or less inflammability, which differs in the facility with which it will take fire, but will infallibly do so if a flame of sufficient intensity is brought into contact with it. Protect it from the flame, and combustion will not take place. The same we constantly find to be the case in epilepsy: remove the exciting cause, and the fits will remain in abeyance; allow the flame to be approximated, and the combustible mixture in your patient's symptom will certainly take fire, the proximity necessary for the purpose constituting the main difference between two different subjects. I have seen this so frequently, that it is one ground why I would specially warn the young practitioner who acts upon these views from placing undue reliance upon the medicinal agents which he prescribes while giving other directions in consonance with the views just detailed. Thus I have again and again found that a continuance of the same active and restless mode of life pursued by a patient at the time of consultation prevented the pharmaceutical appliances from producing a satisfactory result. The patient has then been placed in other circumstances, or has withdrawn from the previous avocations, allowing body and mind the proper rest, and the exciting cause being withdrawn, the nervous system had time to recover its tone, and if not a permanent, still a temporary cure was the result—a cure to all intents and purposes, because it was shown that the epilepsy was under the control of external influences.

A very interesting case which was published many years ago, and has found a place in some works on nervous diseases, may be quoted here as one of the most palpable instances on record of the influence of a physical lesion not affecting the central portions of the nervous system, in producing epilepsy.*

A woman, aged thirty-eight, had been subject to epilepsy for twelve years, when she came under Dr. Short's care. The attack always occurred with a sensation or aura, commencing at the lower part of the gastrocnemii. Dr. Short was so fortunate as to surmise the presence of an irritating agent at the site from which the aura proceeded, plunged a scalpel in, and removed a small hard cartilaginous body, of the size of a large pea, attached to the nerve. From the date of the removal of the neuroma the epilepsy ceased.

A similar case is given in Darwin's "Zoonomia" (p. 329). He states: "I once saw a child, about ten years old, who frequently fell down in convulsions as she was running about in play. On examination, a wart was found on one ankle which was ragged and inflamed, which was cut off, and the fits never recurred."

Similar cases are those in which the introduction of a foreign body into any cavity of the body—as beads into the ear—induces epilepsy, and where *amotâ causâ tollitur effectus;* but in all these instances the obser-

---

* An Epilepsy from an Uncommon Cause. By Dr. Thomas Short, Physician at Sheffield, F.R.S. In Medical Essays and Observations, published by a Society in Edinburgh, fifth edition, vol. iv. 1771.

vation before made applies, that the exciting cause was only capable of inducing the paroxysm because there was some peculiar predisposition in the nervous system, which must be regarded as the "effect defective."

I may append to these remarks a curious instance given by La Motte,* which illustrates how the definite action of an exciting cause may be proved. The importance of avoiding it, where possible, is manifest; but in the present instance this would not have been feasible. La Motte mentions the case of a woman who was eight times pregnant; five of her children were girls, three were boys. Every time she was pregnant with boys she had epileptic seizures, which did not affect her when the child was of the female sex.

---

* Chirurg. Complet., obs. 176, tome ii. p. 422. Also quoted by Esquirol in "Maladies Mentales."

CHAPTER VII.

The pathological anatomy of epilepsy—The import of the lesions found—Dr. Boyd's investigations—Ferrus and Parchappe's inquiries—Wenzel's autopsies—Various lesions—Schroeder van der Kolk's observations—Remarks—Esquirol's inquiries—Further remarks.

IN a disease characterized, as epilepsy is, by symptoms mainly referrible to the brain, or, at all events, invariably associated with evidence of disturbance of the cerebral functions, it is natural that anatomists should have specially searched within the cranium for the organic lesions to which the disease might be attributed. They have been so far successful that, in a large number of patients who have died of long-standing epileptic affections, cerebral lesions have been discovered; but while this is by no means uniformly the case, it is remarkable how little there appears to be of a definite relation between a lesion of a single portion of this complex organ and the convulsions of epilepsy, and how commonly every one of the lesions that have been found associated with epilepsy occur in other patients who have shown no epileptic symptoms. On the other hand, epilepsy is by itself so rarely a fatal disease, that the opportunities of performing post-mortems upon persons who have died of uncomplicated epilepsy, and during or immediately after the fit, are very limited. Hence, in cases of long standing, in which cerebral lesions are met with, it becomes doubly

M

doubtful how we are to interpret these changes; they may not bear any relation to the epilepsy, or if they do, that relation may be one of sequence and not of causation.

If there is this difficulty in regard to cases of epilepsy of an inveterate character, we cannot expect a more satisfactory solution of the phenomena by the post-mortem appearances developed in subjects who have died soon after its first appearance, or in one of the first seizures. We have seen that during the fit itself the effect upon the circulation varies; hence it is reasonable to conclude that some of the post-mortem changes occasionally met with may be the result of the spasm set up, and not the cause of that spasm.

If a railway train runs off its rails, and is precipitated into a river, we do not look upon the broken parapet, or the puddles caused at the side of the bank from the overflow of the river, as the cause of the accident, but we regard them as the unavoidable consequences of the catastrophe; the causes we look for in the state of the rails, of the wheels of the carriages, or in the manner in which the signals have been attended to. In the same way the spasmodic constriction of the vessels of the neck, the disordered respiration, the violent jactitation accompanying epilepsy, may, and undoubtedly do, induce effects, which in the post-mortem must be regarded in the same light as the broken parapet in the instance of the railway accident. But in the case of epilepsy it becomes a matter of yet greater difficulty, than it generally appears to be in the inquiries instituted by the coroner, to determine the real and efficient cause.

While I should be the last to discountenance pathological inquiries, I would express a doubt as to the mystery of epilepsy ever being unravelled by exclusively relying upon the results of cadaveric inspections. It is rather in its vital relations that the disease deserves to be studied than in the dead-house. The interval between the fatal issue and the early stages of the disease is so much longer here than in most of the diseases that ordinarily prove fatal, that an excessive margin is allowed to vague and indefinite surmises; I fear that, until we are better acquainted with the actual nature and mode of transmission of the nervous force, our appreciation of the physical changes accompanying epilepsy will continue removed from the exactness which may be demanded of a scientific doctrine.

I have premised these remarks, in order that the exposition of the post-mortem appearances met with in cases of epilepsy may not be received with too exaggerated an estimate of their value.

We will consider first some of the pathological changes met with in the brain.

But few cases, as I have already said, are fatal early in the disease, or during the epileptic paroxysm. In these an increase in the amount of blood in the vessels of the brain and the meninges, or in the amount of interventricular or sub-arachnoid fluid, has generally been the sum of the post-mortem changes noted by the respective observers. Numerous instances of this kind are quoted by Tissot,* from Drelincourt, Wepfer, Morgagni, Johnstone, and Meckel. But, as Dr. Cooke† justly observes, the fact of the brain being overloaded in the paroxysm, or as its immediate consequence, by no means in itself justifies the inference

---

* Œuvres de M. Tissot, tome vii. 1790.
† A Treatise on Nervous Diseases, vol. ii. part 2. 1823.

that the vessels were in that condition previous to the seizure. At times the vessels are found to have given way, and the consequence of this is an effusion of blood within or upon the brain. One of the earliest instances of this kind on record is quoted by Morgagni,* from Valsalva, in which extravasated blood was found between the dura and pia mater, "besides a quantity of serum everywhere effused; the ventricles were also filled with serum, and in them the plexus choroides had their turgid glandules."

The great variety of lesions met with in epileptics long subject to the disease, and whose death has not been the immediate result of the paroxysm, can scarcely be better demonstrated than by relating successively the post-mortem results in a series of cases found by Dr. Boyd,† who has paid much and careful attention to the subject. In the *Edinburgh Journal* this observer details the post-mortems of six epileptics. Two males presented cerebral wasting; in one female the skull was unusually thick; in one female there was fluid in the brain; in another male the brain was indurated, and presented sharp bony projections from the exterior of the skull; in one male there was congestion of the brain; in two of the females the brain was above the average size, the one firmer, the other softer than natural. In the *Asylum Journal* Dr. Boyd adds the post-mortems of nine other epilep-

---

* The Seats and Causes of Diseases, translated from the Latin of Morgagni, by Benj. Alexander, M.D., vol. i. p. 187. London, 1769.

† See *The Asylum Journal for Medical Science*, April, 1857, p. 377; *Edinburgh Medical Journal*, No. 15, p. 121; Seventh Report of the Somerset County Lunatic Asylum, 1854.

tics, seven of whom were males, and two females. The following are the abridged memoranda of the cases, which show the extreme variation in the local phenomena, even in cases of long duration. We premise, in order that the bearing of the weights may be better understood, that the normal weight of the brain in the male is 46 ounces, in the female 42 ounces.

1. Male, æt. eighteen. Epileptic for thirteen years, with hemiplegia of the right side. There was extensive loss of substance laterally of the anterior lobe of the left cerebral hemisphere; the space filled by fluid was contained in a reticular membrane; the cerebral substance beneath the membrane was smooth, unusually firm, and brownish. Weight of right hemisphere, 21½ ounces; of left, 14¾ ounces.

2. Male, æt. nineteen. Died of typhus. The ridges of the skull were unusually prominent in the temporal fossæ. Weight of the brain, 46¾ ounces.

3. Male, æt. twenty. Died of typhus; epileptic for four years from a fright; an encysted bag, the size of a filbert, over the left cerebral hemisphere, with thinning of the corresponding portion of the skull. Brain otherwise natural; weight, 47¼ ounces.

4. Male, æt. twenty-five. Mania; epileptic for seven years; brought on by a fall attended with fracture of the skull. Brain large, but otherwise natural; weight 49 ounces; skull natural.

5. Male, æt. thirty-three. Recent epilepsy, with mania. Arachnitis; roughness of the lining membrane of fourth ventricle. Brain large; weight 51½ ounces.

6. Male, æt. forty-one. Epileptic from birth; idiotic. Brain small, otherwise natural; weight 38½ ounces. Died of phthisis.

7. Male, æt. forty-eight. Epilepsy for several years,

with mania, caused by close application to business. Congestion to brain; weight, 53½ ounces.

8. Female, æt. thirty-four. Brain large; convolutions flattened; weight, 48 ounces.

9. Female, æt. thirty-eight. Fits from two years; partial paralysis of extremities, which were most convulsed in the fits. Sudden death. Right cerebral hemisphere 2 ounces lighter than left.

I shall again advert to the subject of the weight of the brain, but I draw attention to the fact, that in the above cases the weight was six times above the average, and twice below, and only once normal.

In the Report of the Somerset County Lunatic Asylum for 1854, Dr. Boyd favours us with the results of his observations upon 53 fatal cases of epilepsy which occurred under his care during six years; 30 were males, 23 females, and 12 were also idiots.

The shape and form of the skull presented an abnormity only in one case—a female—in which it was thick behind and the diploe wanting. In nearly one-third there was a difference in the weight between the cerebral hemispheres of from ¼ to 6 ounces. Of the males, 21 were epileptics only, and in them the average weight of the brain was 50·3 ounces; in 9 males who were epileptics and idiots, the average weight of the brain was 46·6 ounces; in 2 only of the latter it was below, but in 6 above the average. The average weight of the *left* cerebral hemisphere in the males was ¾ ounce greater than the right hemisphere. Of the 23 females, 20 were epileptics, and 3 epileptics and idiots; in the 20 epileptics the average weight of the brain was 43·2 ounces, and *right* cerebral hemisphere was slightly heavier than the left: in the two idiots the brain was ¼ ounce less than the average natural weight; and in the third idiot, who died from cerebral apoplexy, which would add to the weight, the brain was 2¾ ounces heavier. One may truly say that the loss of balance so often metaphorically applied to disturbed states of the nervous system, is shown, by the above observations of Dr. Boyd, to bear a literal interpretation. The general increase of weight in the brain of epileptics is a fact which, if confirmed by more extended observation, cannot fail to influence our views with regard to treatment; in speaking of which I shall recur to the point.

Among the authors who have specially attended to the weight of the brain in epilepsy, Ferrus and Parchappe may be mentioned. Parchappe's * observations are not numerous. He gives four cases in which epilepsy and insanity were associated; in three, in which both affections had lasted several years, there was thickening and opacity of the arachnoid, with softening of the middle portions of the cortical layer. In the fourth case the disease had lasted one year only, and the intellectual disturbance had been temporary; there was sudden death from nephritis, but the ventricular arachnoid was rough and thickened, the ventricles contained much serum, and the septum lucidum was very soft. The average weight of the four brains was 1·498 kilogrammes, or 52 ounces 6 drachms.

M. Ferrus, whose researches I am only acquainted with through the medium of M. Parchappe's works, examined the nervous centres in a large number of epileptics; he has almost invariably found hypertrophy of the brain, with increased density, and a brilliant

---

* Recherches sur l'Encéphale, sa Structure, ses Fonctions, et ses Maladies. Par M. Parchappe. Deuxième Mémoire, liv. iii. chap. v.

white colouring of the white substance, together with hypertrophy of the cranium; alterations which M. Ferrus is much disposed to rank as causes of epilepsy. In exceptional cases he has met with other morbid changes, such as softening, tubercle, and hydrocephalus. Together with these observations we may also advert to those of Messrs. Bouchet and Cazauvieilh,* who found a true chronic inflammation of the cerebral tissue to constitute the uniform lesion in epilepsy; a conclusion based upon the post mortems made in eighteen cases of the disease.

As a special opportunity of returning to Dr. Boyd's investigations may not offer, it is convenient at once to add the results which he has obtained by the measurements of the crania of epileptics :—

The average cranial measurements in 30 male epileptics were: circumference 22, transverse measurement from the centre of the external auditory foramen over the head to the other 13·9, and from the root of the nose to the occipital protuberance 13·2 inches; in 29 female epileptics, circumference 21·5, transverse measurement 13·2, antero-posterior 12·9 inches. The average cranial measurements in 14 male epileptics and idiots were: circumference 21·7, transverse 13·7, antero-posterior measurement 13 inches. In 13 male idiots, not epileptics, the average circumference was 21·4; transverse measurement 13·5; antero-posterior measurement 12·8 inches.

Among the observations upon the pathological changes produced in epilepsy few have been made with more searching accuracy than those of Joseph Wenzel.

---

* De l'Epilepsie considerée dans ses Rapports avec l'Aliénation Mentale. Extrait des Archives Générales de Médecine.

They led him to the conclusion that the pituitary body was the only part of the encephalon presenting any uniform lesion. As his inquiries have been much misrepresented and misunderstood, I think it right to give them somewhat in detail. J. Wenzel founded a society in Mayence for the special study of the post-mortem appearances in epilepsy; and unfortunately his death interrupted the progress of the inquiry, but his own results were published posthumously by his brother.

Wenzel's* investigations were made at the beginning

---

* The original edition not being at my disposal when I first entered upon the study of his researches, though I have since examined it for comparison, the extracts are taken from a French translation, entitled, "Observations sur le Cervelet et sur les diverses Parties du Cerveau dans les Epileptiques, par Joseph Wenzel, Docteur-en-Médecine, etc.; publiées après sa mort par son frère et collaborateur Charles Wenzel, Médecin, etc. Traduit de l'Allemand par M. Breton, pp. 217." Paris, 1811.

The title of the original is: "T. Wenzel, Beobachtungen über den Hirnanhang Fallsüchtiger Personen; nach seinem Tode herausgegeben von C. Wenzel." Mainz, 1810. Most English writers who refer to Wenzel have probably made use of the French edition, in which, curiously enough, *Hirnanhang* is given throughout as *cervelet*, which is the common term for cerebellum. Anybody, however, who will be at the pains to look at the French description of the part, will at once understand what is meant; thus, at p. 53, it is stated, "Le cervelet est situé dans la fosse de la selle du Turc;" and a large part of the book is devoted to a description of the sella Turcica and adjoining parts. The error has already been pointed out by Dr. Sims, but I was not at first aware of the correction, and was myself for a time misled by the statements of English authors who had misunderstood Wenzel; it may be right to repeat the warning.

The circumstance deserves to be introduced among the curiosities of literature, not only on account of the glaring error itself, but still more from the ready manner in which it has been endorsed by authors who, with the most ordinary care, might have detected it.

of the present century, and their care and minuteness render them deserving of the especial attention of all interested in the study of the disease under consideration or engaged in the prosecution of any pathological inquiry.

The total number of post-mortems upon epileptics made by Wenzel amounts to twenty; they were not inmates of a lunatic asylum, but patients residing with their friends, or admitted into the general hospital of the town of Mayence on account of other diseases.

The following are some of the main inferences from twenty post-mortems of epileptics, which were conducted with extreme care :—

There are certain deviations or changes of structure of the brain which are found in epileptics, and which are merely coincident with, or consequent upon, the disease; such as variations in the form and size of the convolutions, softening or hardening of the cerebral tissue, accumulation of serum in the lateral ventricles, alterations in the size and consistency of the corpora striata, thalami optici, and corpora quadrigemina. The pituitary body is always found diseased in epileptics, and the morbid condition almost invariably consists in an effusion of lymph, which has become more or less indurated at the point of junction of the two lobes. The pineal gland is also found to be commonly affected; and these two parts are seen to be diseased when no morbid affection can be traced in any other part of the brain. Wenzel is of opinion that the slightest modifications of the pituitary body have the most serious consequences for the animal economy. He regards it as all but certain that the diseased state of the pituitary body in epilepsy is the result of an inflammatory affection; and although he does not go into the details of treatment, he makes his observations the ground of objection to the empirical mode of treatment commonly pursued in epilepsy. Incipient epilepsy, above other remedies, demands, he is inclined to maintain, antiphlogistic treatment, and especially local and general bleeding; or, he asks, is that form of epilepsy alone curable which takes its origin in a distant part of the body, and does not depend upon lesion of the pituitary body? In speaking of the morbid anatomy of epilepsy, Wenzel dwells much upon the necessity of a careful examination of the base of the cranium, as the sella Turcica, the posterior and anterior clinoid processes, are very apt to present some form of malformation in epilepsy. A considerable part of the volume is devoted to this branch of the subject.

An abstract of the cases detailed by Wenzel will, I apprehend, be acceptable to my readers. The results obtained by Wenzel are rendered still more important by the fact that not only his anatomical researches into the brain generally are very extensive and minute, but that in many of the epileptic post-mortems, he examined bodies of the same age, and of persons who had not been epileptic, in order to have a proper standard of comparison. On the other hand, we may not overlook the statement of Rokitansky,* that he has "frequently failed to discover the disease" spoken of by Wenzel "in those who had notoriously suffered from epilepsy and convulsions," and that he had "met with it in other individuals who had been thoroughly healthy."

The question of the bearing of disease of the pitui-

---

* A Manual of Pathological Anatomy, vol. iii. p. 431. Translated by C. H. Moore, Syd. Soc. Ed.

tary body upon epilepsy, like that of Addison's disease to the bronzing of the skin, may therefore still be regarded as a moot point, which deserves further investigation. We subjoin for those curious in such matters the abstracts of the cases given by Joseph Wenzel in a foot-note, because their researches are not much known, nor very accessible.*

---

* Case 1. Male, æt. thirty-eight. Epileptic for four years. Death from phthisis. There was no trace of disease within the cranium, except in the pituitary and pineal bodies; the latter was small, pale grey, and singularly soft; the pituitary body was extremely small; its two lobes were separated from each other by transparent, viscous, yellow matter. There was a corresponding loss of substance in both lobes; the anterior one was throughout of a pale rose colour, and in its middle scarce a trace of the white substance commonly found there remained; the posterior lobe was almost converted into a mere sac.

Case 2. Man, æt. thirty, who had had repeated attacks of epilepsy, and died of typhus. Exhibited no cerebral disorganization, except that the pineal and pituitary bodies were very soft, and the two lobes of the latter separated by a yellowish-brown, viscous, transparent substance.

Case 3. Male, æt. twenty-three. Epileptic from infancy. Post-mortem like the last.

Case 4. Male adult. Affected with epilepsy from infancy. Sudden death. Cerebrum healthy, except pineal and pituitary bodies; the former very soft and small; the latter of "monstrous" size and swollen, on "its upper surface, of a deep red," all the vessels gorged with blood. The whole anterior lobe red inside and outside, "without a trace of white substance within;" the posterior lobe, which is always grey in adults, was of a deep blue, and at the point of junction between the two lobes there was a viscous, yellow, transparent matter.

Case 5. Female, æt. sixty-six. Sudden death. Epileptic for some time. Brain healthy, except pineal gland, which was soft, and pituitary body of a yellowish colour, and nearly equally soft. The posterior lobe filled with pale grey substance, like thick soup; the line of separation between the two lobes was

Although I cannot but admit that the pineal gland and pituitary body are frequently diseased without inducing epilepsy, it is difficult to repudiate the inference

---

filled with yellow, viscous, transparent matter, and both lobes were covered with the same liquid.

Case 6. Male, æt. twenty-three. Died in an attack of epilepsy, having had several previous seizures. Convolutions of brain flattened. The pineal gland small and soft; pituitary body enlarged. A deposit of yellow, transparent matter between the lobes; a vesicle formed on its left side by the investing membrane.

Case 7. Female, æt. fifty. Epileptic for twelve years. The pituitary body contained small round, solid, transparent bodies, of the size of "graines de Séseli," which he could not find in the pituitary body of a man of fifty-five, examined at the same time, who had died of apoplexy.

Case 8. Male, æt. thirty. Epileptic for several years. Died of typhus. Fewer convolutions anteriorly than posteriorly. Corpora striata and optic thalami larger on left than on right side. Pineal gland dried up; anterior lobe of pituitary body externally red; instead of the white spot internally, the part was deep red, verging upon black. Between the lobes there was an empty cavity.

Case 9. Female, æt. twenty-four. Epileptic for several years. Pineal gland much discoloured, very small, and excessively soft. Anterior lobe of pituitary body extraordinarily thick and red. Posterior lobe, which at this age ought to be grey, was white. No yellow matter between the lobes. Brain otherwise healthy.

Case 10. Female, æt. seventeen. Epileptic all her life. Pineal gland very small and soft. Pituitary body enlarged. A white, thick matter, resembling indurated lymph, separated the two lobes. The brain otherwise healthy.

Case 11. A child, æt. two years and nine months. Subject to convulsions from the age of two months. There was much turbid yellow serum under the arachnoid at surface and base of brain; much serum in ventricles. Corpora striata and thalami optici converted into a pulp. Pineal gland very large and pale grey. All the vessels of the pia mater investing the pituitary

of Wenzel altogether, unless we discredit his observations, which it is impossible to do, seeing with what scrupulous and conscientious care they are made and

---

body were gorged with blood. There was an accumulation of turbid and slightly indurated lymph between the two lobes. The anterior lobe was deep red within; and at the point of junction of the two lobes there was in the cellular tissue much semi-fluid lymph.

Case 12. Man, æt. twenty-one. Epileptic for fourteen years. Anterior convolutions of brain very large; the brain healthy, except pineal gland, which was atrophied; the infundibulum, which was as red as if it had been plunged in blood and vermilion; and the pituitary body, the anterior lobe of which, at its upper part, presented on each side a white indurated spot. A light fluid ran out on separating the lobes.

Case 13. Man, æt. twenty-six. Epileptic for several years. Convolutions as in last case. Pineal gland small; upper surface of large lobe of pituitary body hollowed out as in old people. The angles of the anterior lobe of a pale yellow, the remainder of the lobe red-brown. The right angle covered with inequalities. The posterior lobe broadened, and of a pale grey colour, like paste.

Case 14. Female, æt. twenty-four. Epileptic for fifteen years. Pineal gland atrophied; pituitary body presented an extraordinary excavation; the infundibulum much reddened and thickened.

Case 15. Female, æt. fifty-three. Epileptic for some years. No lesion except pineal gland, which was atrophied, and contained two very large calculi. The pituitary body contained in its anterior lobe, which was slightly enlarged, a yellowish-white substance, such as is never found in the healthy state. Posterior lobe pale grey and very soft.

Case 16. Man, æt. nineteen. Epileptic for several years. An effusion on the brain of semi-fluid lymph; no lesion otherwise except pineal gland and pituitary body. The former atrophic. The posterior lobe of latter was hard and very dense, and contained two spots of a brilliant white and cartilaginous appearance, yielding a milky fluid; on removing them a cavity remained,

recorded. Among my memoranda I find such a case as the following, taken by permission from a MS. work submitted to me by Dr. Davy. In a case of pneumo-

---

lined with dense cellular tissue. The interior of the anterior lobe presented two spots of white substance.

Case 17. Man, æt. fifty. Had been epileptic formerly, when he was cured for fifteen years by swallowing a roasted mouse which had been reduced to powder. Return of fits a year previous to death. Vessels on surface of brain gorged with blood. Convolutions of brain widely separated, broad and long. The corpora striata and thalamus opticus of right side almost absent; grey matter very pale ; white matter dirty-white. Pineal gland pale grey, containing much sand. Pituitary body presented a small band a line broad, at the point of junction of the two lobes ; between them was a transparent, clear-yellowish viscous matter, nearly the thickness of a line, a portion of which was indurated. The anterior lobe red at its margins and within.

Case 18. A soldier. Epileptic for several years. Died in a fit. Remarkable narrowing of sella Turcica. Grey substance very pale. Pineal gland enlarged, with a cyst the size of a hempseed on its upper surface, containing a clear yellow fluid. Anterior lobe of pituitary body singularly narrowed; softer than normal; yellow and brownish. The posterior lobe contained a brownish-yellow round spot.

Case 19. Man, æt. twenty-seven. Epileptic for fifteen years. Great contraction of the sella Turcica. Effusion of turbid, whitish serum at the posterior upper surface of the cerebrum ; much limpid serum in the ventricles; the cerebral tissue firm. Pineal gland harder than usual. Pituitary body surrounded by white and indurated pia mater ; surrounding the insertion of the infundibulum the greater part of the surface presented an excavation. Anteriorly was a transparent point the size of a large needle's eye, containing some yellow viscous matter.

Case 20. A child, æt. eleven. Dead of violent convulsions, following soon after smallpox and cynanche. All parts of the cerebrum healthy excepting the pituitary body. At the point of junction of the two lobes was a quantity of deep-yellow, transparent, thick bilious matter, causing a separation of the two lobes to the extent of two lines.

thorax, unaccompanied by epilepsy, the falciform process of the dura mater was found to be cribriform, as if from absorption; the pineal gland was diaphanous, and distended with fluid, the substance of the brain being otherwise healthy.

We have not yet a sufficient knowledge of the functions of the different parts of the brain to decide the question in the affirmative or negative; it is, however, worth remarking that Dr. Todd* has maintained the pituitary body to be connected with the process of absorption, or removal of the effete particles of the brain. This view, if substantiated, might readily be made to harmonize with a theory of epilepsy and Wenzel's observations, a speculation into which I do not, however, wish to enter. It is true that Romberg† has detailed a case of facial neuralgia in which there was no epilepsy, though half the pituitary body was converted into a purplish-brown pultaceous liquid; and he quotes Engel as having met with disease of the pituitary body without epilepsy. Wenzel, however, by no means asserts disease of the pituitary body invariably to be accompanied by epilepsy, but that he has found epilepsy to be invariably accompanied by a morbid condition of the pituitary body. No one appears to have followed up the subject in such a manner as to confirm or refute Wenzel's observations; though, in the case of the two organs specially concerned, it is probable that pathology may contribute more to the elucidation of their functions than physiology. Nor do the observations of Dr. Boyd relative

---

* Cyclopædia of Anatomy and Physiology, vol. iii. p. 703.
† Manual of Nervous Diseases of Man, translated by E. H. Sieveking, M.D., vol. i. p. 43.

to the altered nutrition of the brain in any way contravene those of the German author, since the two sets of lesions might well coexist; and it does not appear that either physician has taken the point of view adopted by the other. The general impairment of the mental faculties that accompanies persistent epilepsy may with all reason be assumed as associated with a different cephalic lesion from that which is immediately connected with the epileptic paroxysm.

Osseous growths from various parts of the dura mater, but especially from the falx cerebri, similar productions within the brain, malformations of the cranium, contraction of the foramen magnum from deposit of new bone, exostosis from the parietal and other bones of the cranium, accumulation of fluid in the ventricles, hydatids in the choroid plexus, tumours of all kinds, the various products of inflammation, circumscribed abscesses—in short, every morbid condition to which the brain and its envelopes are liable, have been met with in epileptic subjects, and have been regarded by authors as the *causa proxima* of the disease. That they may each of them be powerfully exciting causes we willingly admit, except perhaps such formations as Pacchionian bodies or cysts in the choroid plexus, which are scarcely pathological; but we cannot join with those who found upon these changes a distinction between idiopathic and sympathetic epilepsy simply because none of them have as yet been proved to be essential to the disease.

To detail the lesions found by different pathologists in the post-mortems of epilepsy would lead too far, and scarcely avail much to the elucidation of the pathology of the disease. There is scarcely a disease or an injury of the cranium, the dura mater, the cerebrum, the cerebellum, or medulla oblongata, which is not recorded by some author as having been met with in post-mortems of epileptic subjects.* Esquirol at one time thought that the cartilaginous and osseous deposits which he found on the arachnoid of the spinal cord might account for the disease; but apart from other considerations regarding the symptomatology of the disease, the remark holds good here which applies to the individual lesions of the brain above alluded to, that they may occur without inducing epilepsy, and are not, therefore, the essential lesion.

Considering how rarely epilepsy is fatal by itself,† and especially that it possesses so very chronic a character, we should not, from analogy with other morbid conditions, be justified in expecting to find a palpable disorganization except in cases of long standing. The periodical character of the paroxysm, though only, in my opinion, one of the manifestations of the disease, tends to confirm this view. We meet periodicity in many diseases of a type altogether removed from the ordinary character of nervous diseases; but in these there are more marked and continuous manifestations of morbid processes throughout the intervals. The evening exacerbations of febrile affections, of inflammatory disorders, of rheumatism, may be cited in evidence. I am at this moment attending an old lady labouring under a well-marked attack of gout, affecting

---

* The reader would find in Delasiauve's "Traité de l'Epilepsie," under the section *Lésions significatives*, a summary of forty-four necropsies collected from various sources, in which nearly as many different encephalic lesions are reported to have existed.

† Alienist physicians do not hold this opinion, but regard epilepsy as a very fatal disease; it is to be remembered that they rarely see epilepsy except in advanced cases, or complicated with insanity.

the abdominal viscera, in whom the return of pain is as regularly periodical as the attacks of intermittent fever; but, in this as in the other instances cited, the history and the persistent symptoms adequately prove that the whole course of the disease is distinct from what is commonly regarded as an affection of the nervous system. In making this remark I must, however, guard myself against being supposed to advocate the idea that the nervous and vascular systems may, in our estimate of disease, be entirely isolated. I merely contend that the periodic character of the epileptic paroxysm is a ground for not assuming that the disease is dependent upon an organic lesion in the first instance, but that the organic lesions, appreciable to the senses, or rather to our means of investigation, are the secondary effects of the disease.

This would be the place for adverting to the results of Schroeder van der Kolk's[*] inquiries into the pathology of epilepsy. Of all pathologists he alone appears to have approached nearly to what may be regarded as the *causa proxima* of the disease; but then his observations are in a great measure microscopic, and while they mainly apply to the vascular element only, yet want that confirmation by other observers which is necessary to give them general currency.

The learned professor has long regarded the medulla oblongata as the centre from which general reflex

---

[*] Professor Schroeder van der Kolk on the Minute Structure and Functions of the Spinal Cord and Medulla Oblongata, and on the Proximate Cause and Rational Treatment of Epilepsy. Translated from the original by W. D. Moore, A.B., M.B. The New Sydenham Society. London, 1859. See also *Brit. and For. Med.-Chir. Rev.* vol. xix. p. 85, and vol. xxv. p. 73.

movements and convulsions derive their origin, and has therefore sought in it the starting point of epileptic attacks. He holds that even though the primary irritation may be remote—for example, in the intestines—a morbidly elevated sensibility and irritation in the medulla oblongata always form the foundation of such attacks, and render the organ in question more capable of, as it were, discharging itself in involuntary reflex movements. At the time that he first adopted this view he had not made those microscopic investigations of the medulla oblongata in epileptics, which are given in the present work, though he had often observed in epilepsy of long standing that this part of the mesocephale had undergone a process of hardening. More recently he has also met with the opposite condition—viz., one of softening in the same part. In the work before us Schroeder van der Kolk gives the results of fourteen examinations of the medulla oblongata in epileptics. The views we have ourselves advocated with regard to the influence of the circulation in producing the epileptic paroxysm, receive full confirmation from the remark of Van der Kolk, that, to produce epilepsy, no disorganization is necessary, no great change of tissue, but only increased excitability, and commonly augmented determination of blood and chemical change are required. (*Loc. cit.* p. 224.) We must not, therefore, expect to find any marked disorganization displayed by his investigations; but he informs us of changes in the blood-vessels in the part which he considers of especial importance.* The following is an account of the changes observed by Schroeder van der Kolk:—In all dissections of the medulla oblongata in

---

* *Loc. cit.* p. 237.

epileptics, whether death took place in or out of the fit, he met with great redness and vascular tension in the fourth ventricle penetrating into the medulla oblongata, sometimes to a considerable depth. Transverse sections through the whole medulla oblongata, from beneath the pons varolii to the inferior extremity of the corpora olivaria, exhibited the part in the vicinity of the fourth ventricle of a much darker colour, usually containing some more distended vessels, which then ran either in the course of the roots of the hypoglossus into the corpora olivaria, or in the course of the vagus and accessory, or in both. Where the degree of redness was slighter it was commonly confined to the posterior half of the medulla; in most cases, however, this hyperæmia extended into the corpora olivaria, which were often furnished with large blood-vessels. Thus also in the raphe, dilated blood-vessels were almost always visible. After he had discovered the close connexion between the corpora olivaria and the hypoglossal nucleus, the professor happened to find dilated blood-vessels, exactly in this course, in the first epileptic patient whose medulla oblongata he examined microscopically. On measuring the width of the vessels under the microscope, the widest vessels in the course of the hypoglossus were found to amount to 0·230 of a millimetre; in the corpus olivare to 0·305 millimetre; in the vagus, to 0·152 millimetre. He connected this preponderance of the diameter of the capillaries in the course of the hypoglossus over that of the vessels in the track of the vagus with the fact that the patient had invariably bitten his tongue in the fits. On the other hand, he discovered that in a patient who had never bitten his tongue, but in whom the respiration was generally disturbed, the vessels in the course of the vagus were much larger than those in the hypoglossus. These two observations received a general* confirmation from a series of investigations of the same kind, which are presented in the work, to which the reader is referred, in a tabular form. The averages of the different measurements derived from all the cases are given thus; for the sake of comparison we add those made in a healthy medulla:—

Epileptics.	Hypoglossus.	Corpus olivare.	Raphe.	Vagus.
A. Tongue-biters	0·306	0·315	0·355	0·237
B. Not biters	0·210	0·217	0·300	0·348
Difference	+0·096A.	+0·098A.	+0·055A.	+0·111B.
Healthy Med. Oblong.	0·097	0·052	0·148	0·064

The experiments of Kussmaul and Tenner have a bearing upon the observations of the Dutch professor; but as they are exclusively physiological, and are performed rather with a view to a synthetical production of epilepsy than an analytical examination of its postmortem results, I postpone giving an account of them till I discuss the theory of the disease.

I have, in my opening remarks, already dwelt upon the necessity of not regarding the paroxysm or fit in any other light than as a part of the entire morbid condition; but the symptoms traceable during the interval are so faint, and require so much careful and minute inquiry to be elicited, that they are apt to be overlooked by those who expect as palpable deviations

---

* I use the term advisedly, because one case is given of a tongue-biter in which the vessels in the hypoglossus were considerably larger than those in the vagus.

VALUE OF POST-MORTEM RESULTS. 183

from health here as in those disorders which are universally regarded as depending upon blood disease. Still it is by their analysis that we may more confidently hope to attain to a just estimate of the organic lesions accompanying epilepsy, than by the most careful examination of the explosive phenomena of the seizure taken alone.

In speaking of the value of the various circumstances influencing the causation of epilepsy, I observed that they occupied a different relation; some acting more immediately by predisposing the body to the occurrence of the paroxysm, when some peculiar stimulus is applied; others, occupying the place of that stimulus, and coming under the denomination of exciting causes. There is some support to this arrangement to be found in pathological anatomy; and it is doubtless important in estimating the value of the relative post-mortem phenomena, that some such rule of measure should be applied; otherwise it would be difficult to understand why, for instance, a spiculum of bone pressing upon the brain, a sword's point, or a musket-ball imbedded in it, may in one person produce epilepsy, while similar lesions in another are not followed by any such phenomena.

In judging of the post-mortem appearances of epileptic subjects, then, we must seek to find or distinguish three things: first, the state of the brain which may give rise to epilepsy or permit its occurrence on the approach of a given excitant; second, the state of the brain and its envelopes, which may be regarded as the excitant; and, thirdly, the condition of the brain or individual portions, which are affected in consequence of the derangement of the circulation and the nutrition of the parts immediately resulting from the epileptic seizure.*

I agree with those who think that every vital act is accompanied by a change in the organism, and that therefore every morbid condition must necessarily be associated with some physical lesion. Our means of investigation are much too coarse as yet to measure the lesion in the majority of cases of functional derangement, and however much we may advance in our mode of inquiry, we shall never bridge over the distance between death and life. There may be numerous organic lesions accompanying so-called functional derangement during life, which we may attain to measuring by secretions and excretions, by dynamometers, galvanometers, and æsthesiometers, but which the cessation of life places absolutely beyond the reach of those methods of estimation, even when perfected to the utmost, simply because the conditions are absolutely altered. While, therefore, we are to keep in mind the possibility of an appreciable change in the encephalon being necessary that the reaction to a certain stimulus may be shown by the epileptic paroxysm, we must also remember that it is perfectly possible that that change may be of a character to disappear entirely with the cessation of life. These remarks would apply to other diseases as well, but to none so forcibly

---

* I leave these remarks as they stood in the first edition; but to render my meaning more precise, I may add that I regard the great majority of lesions discoverable in the brain in epileptics in the same light as any other eccentric cause; the influence they exert in the production of the paroxysm being conveyed by reflex to the mesocephale or such parts of it as are immediately concerned in the fit, in the same manner as the irritation that proceeds from a morbid ovary or an irritated median nerve.

as to those accompanied by what are termed nervous symptoms.

The poverty of our knowledge concerning the functions of the nervous system and the manner in which they are carried out, justifies us in setting aside altogether the consideration of the physical condition of the brain which predisposes to epilepsy. What else I may have to say on this subject, I reserve until I discuss the theory of the disease.

The question then suggests itself, to which of the two remaining categories the lesions belong which are most frequently met with in conjunction with epilepsy; are they to be viewed in the light of exciting causes, or as the results of the concomitant affections of the disease? The question is placed here, because it appears to be one deserving an answer, though a sufficiently exact and satisfactory response may yet be wanting.

I have put forward so prominently the observations of Dr. Boyd and Professor Wenzel, because they in no way contradict one another, and both tend to show a peculiar chronic malnutrition of the brain. The hypertrophy demonstrated by Dr. Boyd has not been inquired into microscopically or chemically, therefore we do not know whether it depends on a new deposit of a heterogeneous character, or whether it is the product of inflammation. Dr. Wenzel's observations of changes in the pituitary body point more definitely to inflammation, because we find both the early stages of this process as well as the suppurative form. It is perfectly conceivable that the hypertrophic condition might have coexisted in Wenzel's cases, had the balance been employed; it is equally possible that, had the pituitary body been examined by Dr. Boyd, he might have met with similar alterations as the former inquirer. I do not like dealing in suppositions, but they are offered in the present instance merely to show that there is no necessary contradiction between two excellent observers. In all their cases the disease had been of long standing, and I incline to the opinion that a morbid process once having been set up by the peculiarity of the spasm and the violent perturbation of the circulation, this has been maintained on the principle that the weakest organ invariably suffers most in any disease that may be set up. The epileptic paroxysm once having occurred, say by an irritation conveyed to the medulla oblongata or the mesocephale, and having initiated a peculiar process in a part of, or in the entire brain, the process might be justly assumed to react upon the epileptic paroxysm, causing its re-excitement; the epilepsy and the organic lesion would then come to stand in the relation of the τύπος ἀντίτυπος of the Delphian oracle.

We shall see that these points exercise a material influence upon the treatment of the disease; and though morbid anatomy by no means affords us all the replies to our queries that we should desire, still the facts it does supply are sufficiently broad and intelligible, and present a sufficient agreement, to enable us to fix certain general principles of treatment in consonance with them.

I have spoken of the observations of Esquirol relative to the presence of cartilaginous and osseous formations on the spinal arachnoid. That such formations may influence the occurrence of epilepsy I raise no question about; but our present knowledge of the functions of the nervous system and the phenomena of the disease in question, forbid our regarding any affection of the spinal cord, or of the nervous system extraneous to the cephalic centre, in any other light than that of an exciting influence. In twelve epileptic women Esquirol found the spinal meninges injected once, and twice of a greyish colour; nine times they presented more or less concretions scattered over the external surface of the spinal arachnoid; the concretions were of a lenticular form, from one to two lines by one line, and they were cartilaginous or bony; while at the same time M. Metivié found similar concretions in the spinal column of two epileptic children. Esquirol found the spinal cord, especially the lumbar portion, softened four times.

In the first edition I remarked that, considering the relation of epilepsy to spasm in the range of the respiratory tract of nerves, it was remarkable that the changes found in post-mortem examinations should so rarely be traced to the medulla oblongata of epilepsy, and that irritation is propagated to it can scarcely be doubted by any one who observes the phenomena of the seizure, whether or not he adopt Dr. Marshall Hall's tenets; and also that, considering the facility of examining the medulla oblongata, and the rarity with which lesions are discovered in it, the view that it is essentially concerned in the propagation of the epileptic paroxysm receives a strong negative. The researches of Schroeder van der Kolk call for a modification of the view presented in this passage, but even they show how comparatively trifling the lesion is which an attentive pathologist can detect in the part. It is only just to Dr. Copland* to state that he especially dwells upon irritation of the medulla oblongata in epilepsy in an

---

* Dictionary of Practical Medicine, art. Epilepsy.

article apparently written long before Van der Kolk's investigations were commenced. It was, however, reserved for the latter to give pathological facts in favour of this view.

Among the lesions not immediately connected with the nervous centres, which at times possess a manifest relation to epilepsy, there are none that can be fairly regarded in any other light than as accidental exciting causes. Injuries to the nerves by wounds or by the spontaneous growth of tumours belong to this class; degenerative disease of the kidneys, the chief depurators of the blood, equally deserves mention here; but, however important to attend to the removal of such morbid conditions to secure the patient's recovery from the epilepsy, they are not essential accompaniments of epilepsy. We have seen (p. 117, seqq.) how very rarely albuminuria, the prominent symptom of degenerative renal disease, accompanies ordinary epilepsy. On the other hand, puerperal convulsions have been shown to be commonly associated with it; and wherever we find albuminuria we must be on the watch for the possible supervention of epileptiform seizures; but a majority of cases of Bright's disease of the kidneys, whether in the form of the large mottled fatty kidneys, or the granular and contracted variety, pass off without the occurrence of epilepsy. Again, I have seen an arrest of albuminuria followed not only by an arrest and apparent cure of severe dropsy, but by complete arrest of severe epileptiform convulsions, which had supervened during the existence of the albuminuria, and which had every appearance of a tendency to death.

CHAPTER VIII.

The theory of epilepsy—General remarks—Its proximate cause —Esquirol's classification—Prichard's view—General remarks—Sir A. Cooper's experiments—Schroeder van der Kolk's theory—Brown-Séquard's experiments and doctrine—Kussmaul and Tenner's views and experiments—Remarks—The influence of habit—Relation of epilepsy to kindred diseases—Metastasis in epilepsy—State of the blood in epilepsy—Dr. Handfield Jones' case and remarks—Acute and chronic epilepsy—Cases in illustration of author's remarks.

If we apply a lighted taper to a muslin curtain, the boarding of a wooden hut, or solid masonry of a church, the effect will vary with the greater or less inflammability of the different substances. The curtain will speedily take fire and flare away; the planks may be scorched, but will probably not inflame; while the stones will show no traces of the influence of a destructive agent which the first shower will not wash away. In the first two instances there is a possibility of ignition, in the third it is not possible. Mankind vary similarly in their tendency to nervous disorders generally, and to epilepsy especially; some are utterly insusceptible to influences that may produce them; others, like the wood and the muslin, are more or less impressionable. But wherever the disease occurs, it is essentially the same disease; the same symptoms characterize it; it follows the same course, and, unless checked, leads ultimately to the same results.

In regard to treatment, it is doubtless of importance that we should discriminate between the various causes that induce the malady, because in this case, as in the whole domain of pathology, the remedies to be applied will depend very much upon the circumstances connected with and immediately preceding the outbreak; but though the removal of those circumstances may arrest or mitigate the disease after it has shown itself, they do not constitute the disease; just as the lighted taper, in the instance above quoted, does not constitute the conflagration which may ensue if the muslin or the wood is ignited by its agency.

Owing, as I think, to a misapprehension of the true nature of this relation, authors have been at great pains to establish a variety of classifications of epilepsy; which, however apparently differing from one another, agree in this, that they make the various exciting causes of the disease the basis of numerous divisions. No injury would accrue if the only consequence were an indication as to the mode of treatment; but the system of classification, which is based upon the exciting causes, induces a species of fatalism, amounting to this, that if the exciting cause cannot be traced, we are unable to control the disease. I conceive this proceeding to be unjust to medical science, and equally discouraging to the medical man and the patient.

Esquirol divides epilepsy into essential, sympathetic, and symptomatic epilepsy; Dr. Cooke classifies it as idiopathic or symptomatic; Sauvages gives no less than fourteen divisions, which are all treated by him as occupying the same nosological rank; they are—
1. Epilepsia plethorica; 2. Cachectica; 3. Stomachica; 4. Uterina; 5. Simulata; 6. Pedi-symptomatica; 7. A pathemate; 8. Sympathetica; 9. Febricosa; 10. A

dolore; 11. Exanthematica; 12. Syphilitica; 13. Traumatica; 14. Rachialgica.

Recent British writers generally adhere to the division into the idiopathic or centric, and symptomatic or eccentric form of epilepsy; the practical result of which is that only those cases are regarded as curable which are symptomatic, and, by reasoning in a circle, all cases that are cured are set down as symptomatic or non-essential.*

Not all writers, however, have adopted the views adverted to in the preceding remarks. If I understand Dr. Prichard rightly, he speaks of epilepsy only as one disease, which presents certain variations in its phenomena, but which are all closely allied to one another, and based upon the same essential morbid condition. The writer, however, who has most clearly asserted the essentiality of all cases of epilepsy is Georget,† who says:

---

* No one has put this more strongly than Dr. Russell Reynolds, whose important work on Diseases of the Nervous System deserves to be studied by all interested in this department of medicine. I must beg leave to differ from him on this point, but as his views are those adopted by many writers, I quote the following passage: "If we can succeed in distributing all the cases hitherto known as epilepsy among the several classes of better-defined diseases, we ought to reject the term epilepsy from our nosology; but if we cannot accomplish this distribution, and are compelled to recognise the existence of many, or even of a few, cases distinct from any more general condition of systemic or local disease, then we must employ the term (epilepsy) in a restricted sense, implying only those cases which, in the present state of medical science, are irreducible."—*The Diagnosis of Diseases of the Brain, Spinal Cord, Nerves, and their Appendages,* by J. Russell Reynolds, M.D., &c., p. 174. London, 1855.

† De la Physiologie du Système Nerveux et Spécialement du Cerveau, tome ii. p. 363. Par M. Georget, D.M. Paris, 1821.

" Si, pénétré des principes de physiologie et d'anatomie pathologique, l'on veut observer sans prévention l'épilepsie, l'on sera convaincu que cette maladie est une affection idiopathique du cerveau, et l'on se rangera de l'avis de Pison, Willis, et De Moor."

Our knowledge of the physical changes of the brain is not sufficiently advanced as yet to speak positively of the exact pathological derangement that accompanies or induces epilepsy. But the analysis of the symptoms of the disease, based upon the knowledge we do possess of nervous physiology; the study of the consequences of epilepsy as shown in a large majority of epileptic post-mortems; the impossibility of rigidly carrying out the distinction between essential and non-essential, idiopathic or symptomatic epilepsy,—all justify us in discarding such an arrangement. Besides, however much we may value the results afforded by pathological anatomy, we must bear in mind that many of the observations made with regard to epilepsy have indicated lesions the exact bearing of which we are scarcely able to interpret, except in so far as they confirm the fact of the cephalic centre being deeply implicated. The observations of the Wenzels and of Dr. Boyd, interesting as they are, at present stand almost isolated, and require confirmation; though both series of inquiries prove how unscientific it is to draw positive conclusions solely from the appearances presented on a casual and superficial examination of the brain. I would ask, in how many post-mortems the pineal or pituitary bodies are ever specially examined, and whether they are not generally passed over simply because we do not know that they preside over a special function? How often is the specific gravity of one or the other part of the brain taken, or how often is one hemisphere weighed against the other? Moreover, though our knowledge of the nutritive processes in the brain has of late years been much extended, especially by the use of the microscope, it is impossible to deny that, with the more recondite changes that take place in the ganglionic and tubular matter of the brain and their relative relations, we are but very imperfectly acquainted.

With the exception of Esquirol (whose observations of the affection in the spinal meninges in epilepsy have been previously recorded), and Marshall Hall, all pathologists are unanimous as to the fact that the parts within the cranium are the organs immediately involved in the epileptic seizure. Here, as in all nervous affections, we may apply what Romberg, in poetic imagery, has said of the symptom of pain, which he terms, " the prayer of the nerve for healthy blood." Spasm no less deserves this designation; but while we must not lose sight of the agency of the blood, we are bound to localize the disease as much as possible, in order that we may determine the means of attacking the immediate derangement which induces the spasm.

It is manifest that, in estimating post-mortem lesions of any kind, we must carefully distinguish between the results of long-standing disease and the appearances accompanying its earliest manifestations. The former, seen alone, may not, except by analogy, enable us to form any conception of the latter. Who that had only seen pneumonia in the state of dense hepatization would be able to appreciate the previous phases of the disease? Who that regarded only the stone in the bladder without reference to the dyspepsia or renal disturbance that had preceded, would be able to account for its formation, and take steps to anticipate its formation in other cases? In each of these instances we may receive indications and suggestions during our post-mortem inquiries; but they will not and cannot, without physiological knowledge—without an acquaintance with functional disturbance manifested and ascertained during life, afford any very valuable information. In the same way, and yet more in regard to diseases of that part of the system which may be regarded as the special bearer of life, and of life in its highest sense, our researches into the value of tangible cadaveric changes must receive their stamp from our knowledge of the vital changes of the functions. The organic lesions in the disease in question, as far as we now know, are only the result of long-continued diseased action; and we are not even in a position positively to affirm which part of the complex organ in which they are found is the part primarily involved; and yet we are justified by all our previous knowledge of general physiology, as well as of the physiology of the nervous centres, in assuming that each portion has a definite function, and that a disease of so peculiar a character as epilepsy is accompanied by a derangement of function primarily of one part of the cephalic centre.

When we wish to form an estimate of a disease we naturally turn our special attention to the most prominent and most tangible symptoms. At the outset I observed—and I think it a subject of such importance that it will bear frequent repetition—that the epileptic fit is only a part of the total disease; still the study of this part is necessary, and shows most markedly what part of the system is specially affected. The sudden and entire loss of consciousness alone would indicate an affection of the encephalon, and especially of the sensory ganglia; the only organ which must be excluded in reference to this point is the heart, as syncope might be mistaken for the insensibility of epilepsy; the state of the heart and the pulse, the spasmodic jactitations, however, prevent an error in this direction. The control of the brain being withdrawn, the spinal system acquires a preponderating action, as shown in the spasmodic action of various muscular terminations of spinal nerves. Whether the irritation proceeds from without or acts directly upon the brain, appears to be of great moment with reference to the mode of treatment to be pursued, but does not affect the theory of the proximate cause. In the phenomena of cerebral reflex, certain molecular changes take place within the brain on the application of a stimulus conveyed through the sensations; and we know equally that morbid impressions conveyed to the brain from without may induce organic changes within the organ. There is an objection to bringing forward isolated cases in proof of a general law, but I may quote the well-known history of the soldier given by Lallemand,* in illustration of my present meaning: " A soldier was operated upon for aneurism of the right axillary artery. In applying the ligature the nerve was enclosed, cerebral symptoms followed on the seventh day, and death ensued on the eighth. The post-mortem showed an abscess in the left posterior cerebral lobe. The case is also of interest as affording proof of the uniformity of the law of crucial conduction." Here the external injury was the probable and immediate cause of the cerebral disease; and it is reasonable to assume that, in a similar manner, external exciting causes operate in

---

* Recherches Anatomico-Pathologiques sur l'Encéphale, vol. i. p. 123; and Manual of Pathological Anatomy, by Drs. Jones and Sieveking, p. 255.

the production of epilepsy by causing a specific irritation of a definite portion of the encephalon. Dr. Carpenter * regards "the sensory ganglia as the primary seat of that combination of loss of sensibility with convulsive movements which constitutes epilepsy;" and he subsequently observes—"The disease cannot be fairly attributed to those obvious lesions of structure which are sometimes coincident with it, and which, as Dr. Todd has justly remarked, are rather signs of altered nutrition, brought on by any cause which creates frequent disturbance of the actions of the brain, than the causes of that disturbance."

That a change in the balance of the circulation has a material influence in the production of epilepsy I think can scarcely be doubted. This is particularly manifest when change of position causes the fits, as in passing from the erect to the recumbent, from the recumbent to the erect posture, in stooping and the like. Where attacks come on in the morning after rising I have little doubt that the passage of the subarachnoid fluid from the brain into the spinal canal may often be an efficient cause, from the more syncopal character of the fits occurring under these circumstances.

Sir Astley Cooper's experiments † afford strong evidence in favour of this view.

He states that, having tied the carotids in a rabbit, "Respiration was somewhat quickened and the heart's

---

* Principles of Human Physiology, fifth edition, p. 673. London, 1855.

† Some Experiments and Observations on Tying the Carotid and Vertebral Arteries, by Sir Astley Cooper, Bart., Guy's Hospital Reports, vol. i. p. 457.

action increased, but no other effect produced. In five minutes the vertebral arteries were compressed with the thumb, the trachea being completely excluded. Respiration almost directly stopped; convulsive struggles succeeded; the animal lost its consciousness, and appeared dead. The pressure was removed, and it recovered with a convulsive inspiration. It laid upon its side, making violent convulsive efforts, breathed laboriously, and its heart beat rapidly. In two hours it had recovered, but its respiration was laborious. The vertebrals were compressed a second time. Respiration stopped, then succeeded convulsive struggles, loss of motion, and apparent death. When let loose its natural functions returned with a loud inspiration, and with breathing excessively laboured. In four hours it was moving about, and ate some greens. In five hours the vertebral arteries were compressed a third time, and with the same effect. In seven hours it was cleaning its face with its paws. In nine hours the vertebral arteries were compressed for the fourth time, and with the same effect upon the respiration. After thirteen hours it was lively. In twenty-four hours the vertebrals were compressed for the fifth time, with the same result—viz., suspended respiration, convulsions, loss of motion and consciousness. After forty-eight hours, for the sixth time the same results were obtained by pressure. Thus it appears, if the carotids are tied, that simple compression of the vertebrals puts an entire stop to the functions of the brain." The experiment was reversed—the vertebrals tied and the carotids compressed, with similar result. Tying the vertebrals caused the breathing to become laborious; the animal's right ear fell, and the right fore-leg was partially paralysed. In five hours it ran about. The following day, when the carotids were compressed, it fell on its side, losing all sensation and volition, and recovered on withdrawal of pressure. The same results were obtained repeatedly. When both vertebral and carotid arteries were tied at the same time, " the animal breathed no more; but there were thirteen to fourteen convulsive contractions of the diaphragm and convulsions of the hinder extremities, and the animal ceased to exist."

The proximate cause of epilepsy is still *sub judice*. But since the preceding remarks were first published numerous inquirers have advanced the question considerably nearer to a solution. The general tendency of the doctrines inculcated by them appears to me to support the views I have advocated. I allude more especially to the researches of Van der Kolk, Brown-Séquard, and Kussmaul and Tenner. Their works are before the public, but a brief summary of their inquiries may not be unacceptable to my readers, as no argument on the subject of epilepsy would be complete without due reference to them. Schroeder van der Kolk, after detailing his discoveries of the peculiar state of the vessels in the medulla oblongata in epilepsy, of which I have already spoken, makes the following observations on the intimate nature of the disease, which I quote literally in order that I may not in any way misrepresent him:—

" I think we have sufficient reason to conclude that the first cause of epilepsy consists in an exalted sensibility and excitability of the medulla oblongata, rendering the latter liable to discharge itself, on the application of several irritants which excite it, in involuntary reflex movements. This irritation may either be external (irritation of the trigeminus), an irritated condition of the brain, or, as is still more frequent, it may proceed from irritants in the intestines. In children worms in the intestines, acidity, a torpid state of the bowels, &c., are among the most common causes; in adults there may be irritation of the intestines, particularly of the mucous membranes, constipation, and prolongations of the colon connected therewith, but above all onanism, which acts so very much on the medulla oblongata, and must be regarded as a very frequent cause of epilepsy. Amenorrhœa, chlorosis, plethora of the uterus, hysteria, &c., are also to be enumerated.

"In the commencement there is still only exalted sensibility. If this can be removed or moderated, the epilepsy gives way of itself, especially if the sensibility is not renewed by remote causes. But if the disease has already lasted long, organic vascular dilatation takes place in the medulla, the consequence being that too much blood is supplied, and the ganglionic groups are too strongly irritated—too quickly overcharged. Every attack then becomes a renewed cause of a subsequent attack, as the vascular dilatation is afresh promoted by every fit. Lastly, increased exudation of albumen ensues from the now constantly distended vessels, whose walls at the same time become thickened, producing increased hardness of the medulla, subsequently passing into fatty degeneration and softening, and rendering the patient incurable."

Whether we regard the medulla oblongata as the chief seat of the malady or not, it is impossible not to see in the above precise and logical account the nearest approach that has yet been achieved to a rational theory of the disease. It harmonizes certain physiological principles with pathological data observed by the author. To a certain extent I adopt it, but I cannot at present pin my faith to it entirely; for this reason, that it does not take sufficiently into account the great fact in epilepsy, that the leading symptom is not convulsion, but unconsciousness. In the minor attacks, a passing unconsciousness is constantly recurring without a semblance of disturbance of the respiratory function, or of those functions over which the hypoglossus presides. In the full seizure the convulsions do not set in till some time after the unconsciousness. This is essentially the primary feature; as the first stage of the fit, which is characterized by a deadly pallor, passes into the second one of livid, venous congestion, the convulsions commence; there is difficult respiration, froth at the mouth, bitten tongue, gurgling and jactitation of the limbs, and contortions of the face and trunk. Besides, in a large number of cases of genuine epilepsy, there is no serious disturbance of the respiratory function throughout; in the variety that has been termed epilepsia syncopalis, this is particularly the case. On the other hand, we occasionally meet with instances in which spinal symptoms predominate, and where a spasmodic movement of the extremities ushers in an attack—movements that at times by themselves constitute a *petit mal*, that may be arrested by a ligature or a blister, or that may be followed by complete unconsciousness; this symptom, however, again constituting the first evidence of the actual seizure. Or are we required to look upon the initiatory scream as the characteristic feature? The answer is, that it cannot merit such distinction, for the reason that it only occurs in some patients, and by no means in all; and whether it takes place or not, it does not alter the sequence of the other phenomena as above detailed. Besides, the contraction of the pupil, which I believe commonly to occur at the onset of the fit, cannot be explained on the view of Schroeder van der Kolk, that the *causa proxima* resides at the origin of the vagus and hypoglossus, but would point to the origin of the third and fifth pairs.

Dr. Brown-Séquard* holds that the seat of epilepsy is very variable, and that it may be the result of an excitation of the whole cerebro-spinal axis. He considers that there are three distinct elements for the production of a fit. 1. Increase of the *force* of the reflex property. 2. Increase of the *excitability* of this property. 3. An excitation of a special nature, or a very violent one. For his explanation of these terms, and the application of his theory to the various phenomena, I must refer the reader to his book; I wish, however, to advert to the interesting manner in which Dr. Brown-Séquard applied the discovery of Claude Bernard, relative to the influence of the sympathetic on the blood-vessels, to the earliest symptoms of a seizure. Having found that irritation of the sympathetic by galvanism induced a contraction of the blood-vessels, he came to the conclusion that the first stage of the paroxysm, which is characterized by pallor, is due to irritation of the sympathetic, which supplies the face, and that at the same time branches of the same or other nerves going to the blood-vessels of the brain are irritated. Hence a contraction occurs in these blood-vessels, and particularly in the small arteries; this contraction expels the blood, and the brain at

---

* Researches on Epilepsy. Boston, 1857, pp. 26 et seqq.

once loses its functions, just as it does in a complete syncope. In consequence of the impairment of the respiratory process, the blood becomes unduly charged with carbonic acid, and thus endowed with the power of exciting convulsions. Dr. Brown-Séquard is of opinion that the contraction of the small vessels in the brain causes an accumulation of blood at the base of the brain and in the spinal cord, so as to favour the irritation of the undecarbonized blood upon the medulla pons and tubercula quadrigemina.

It is impossible for me to go fully into the interesting researches of Dr. Brown-Séquard here, but I cannot pass to the results obtained by Messrs. Kussmaul and Tenner without adverting to the remarkable production in the guinea-pig of epileptiform convulsious by operative procedure. Assuming the proof of the existence of epilepsy in these cases to be absolute, they prove how an exalted condition of the seat of the epileptic "affection" may be induced by an injury to a distant portion of the nervous system, and how, on the application of a given irritant within a certain range of nerves, we may, under these circumstances, voluntarily excite the paroxysm.

Dr. Brown-Séquard[*] has found that transverse section of part of the whole spinal cord, and puncture of the same (especially in the region between the seventh and eighth dorsal and third lumbar vertebra), in guinea pigs and other animals, are followed by epileptiform attacks. They begin during the third or fourth week after the infliction of the injury, and consist, according to the extent

---

[*] Researches on Epilepsy. Its Artificial Production in Animals. and its Etiology and Treatment in Man. Boston, 1857.

of the injury, in brief spasm of the muscles of the face and neck, including those of the eye and tongue; the head is drawn to one shoulder, and the mouth opened by the spasm of some muscles of the neck. After a time the fits are more complete, the convulsions becoming universal, with exception of the parts that are paralysed. The convulsions may come on spontaneously or be excited, and it is found that the irritation capable of producing them must be applied to certain parts of the face and neck, the side to which it is directed corresponding with that of the injury. This part is described by the author as being circumscribed by the following lines—" one uniting the ear to the eye, a second from the eye to the middle of the length of the inferior maxillary bone, a third which unites the inferior extremity of the second line to the angle of the inferior jaw, and a fourth which forms half a circle and goes from this angle to the ear, and the convexity of which approaches the shoulder." If a portion of the skin is detached—so, however, that connexion is maintained by the infra-orbital nerve—convulsions may still be induced by irritation of this piece of integument. The author believes that pain has little to do with the production of the fits, but that they result from reflex. The spasmodic action, the falling of the animal, the drawing of the head to one side, the irregular respiration, the expulsion of fæces and urine, closely resemble epilepsy, as I can myself testify, having had the advantage of witnessing some of Dr. Brown-Séquard's experiments. Still, it is doubtful whether there is total insensibility, and Dr. Brown-Séquard himself appears to feel some uncertainty as to whether the phenomena in question are identical with epilepsy, as he remarks that, if they are not truly epileptic, they are at least epileptiform, while he enumerates three points in which they differ from epilepsy as observed in man. It is not to be understood that he regards these experiments as proving the *causa proxima* of epilepsy to reside in the spinal cord alone; on the contrary, he distinctly states that the seat of epilepsy seems to be "in the part of the brain where reside the faculties of perception and volition, and in the part of the cerebro-spinal axis endowed with reflex faculty; but," he continues, " whatever may be thought on this subject, it seems quite certain from facts observed in man and animals that epilepsy may be produced by various alterations of the encephalon of the spinal cord, and of a great many nerves. In other words, the peculiar disturbance of the cerebro-spinal axis which constitutes epilepsy may be generated by alterations of various parts of this nervous axis and by many nerves."

I cannot dismiss this valuable inquiry without pointing out one peculiarity which the author does not advert to in the enumeration of the features of the artificial disease distinct from those of acknowledged epilepsy in man; it is the remarkable hyperæsthesia that prevails in the animals subjected to experiment. I have seen, I think, three cases,* in which something analogous appeared to exist in the human being; but I would suggest, with all deference, whether the persistence of sensibility, admitted by the author, and the remarkable sensitiveness to external impressions resulting from the operation, do not establish an

---

* They will be adverted to when I speak of the local treatment applicable to the disease.

analogy between the phenomena observed, with hydrophobia almost as much as with epilepsy.*

Kussmaul and Tenner† conclude from their experiments, that a circumscribed anatomical alteration of the brain must not be regarded as the *proximate* cause of epileptic attacks; it is only in microscopic alterations of the brain that the cause of epileptic affections resides. Hence, they teach, as I have already pointed out, that

---

* Although not disposed to adopt Dr. Watson's views with regard to the absence of cerebral congestion in the epileptic paroxysm, I think the following remarks explain and admirably sum up our knowledge of the true relation of the brain and spinal cord in epilepsy:—" There are good reasons for believing that the change, whatever it is, which is the immediate precursor and cause of the epileptic fit, may sometimes originate in the spinal cord, and thence extend to the brain, and sometimes originate in the brain and communicate itself to the spinal cord. Dr. Marshall Hall's doctrine that all convulsive diseases are diseases of the spinal marrow, cannot properly be applied to this convulsive disease of epilepsy. It is true that the spinal cord is concerned whenever there is convulsion, but it is concerned in every *voluntary* movement also, through the instrumentality of the brain itself; and it may be, and often is, irregularly influenced by a disordered and unnatural state of the brain. Tetanus may be fairly regarded as a disease of the cord and its proper appendages; the spasms arise and reach their height, while the powers of thought and sensation are undisturbed, and while the volition remains, although the morbid condition of the cord renders it ineffectual. In epilepsy these cerebral functions are always implicated. There is *always* a loss of consciousness, and in the epileptic vertigo, the *petit mal*, there is frequently a suspension of consciousness only, *without any convulsion at all.* The brain, therefore, we must consider to be essentially concerned in this disorder.—*Lectures on the Principles and Practice of Physic*, vol. i. p. 620. By Thomas Watson, M.D. 1843.

† Selected Monographs. New Syd. Soc. London, 1859, pp. 85, seqq.

tubercle of the brain, cicatrix of the brain, of the spinal cord, or of a cutaneous nerve, can only be regarded as remote causes of epilepsy; and should visible alterations occur in the brain or other parts of the body during eclampsia and epilepsy, they must be regarded as nothing else than predisposing influences. Again, they point out that the proximate cause of the attacks cannot be one of long duration, but an alteration merely of a temporary kind; because it must be quickly developed to its full extent, pass during the attack through its different phases, and when the latter are over, cease completely or nearly so.

Valuable as the work of these gentlemen is, we must bear in mind that it is devoted to the investigation of the phenomena resulting from a single mode of influencing the nutrition of the brain—viz. by the abstraction of arterial blood. This they in fact themselves admit. It would not therefore be legitimate to take the results they arrive at as illustrative of more than a portion of the pathology of epilepsy.

I extract the following passages, bearing specially upon epilepsy, from the summary given by themselves at the end of their treatise:—

" The convulsions appearing in profuse hæmorrhage of warm-blooded animals (including man), resemble those observed in epilepsy.

When the brain is suddenly deprived of its red blood, convulsions ensue of the same description as those occurring subsequent to ligature of the great arteries of the neck.

Epileptic convulsions are likewise brought on when the arterial blood rapidly assumes a venous character—as, for example, when a ligature is applied to the trachea.

It is highly probable that in these cases the attack of spasms depends upon the suddenly interrupted nutrition of the brain; it is not caused by the altered pressure which the brain undergoes.

Epileptic convulsions in hæmorrhage do not proceed from the spinal cord, nor from the cerebrum. Their central seat is to be sought for in the excitable districts of the brain lying behind the thalami optici.

Anæmia of those parts of the brain situated before the crura cerebri produce unconsciousness, insensibility, and paralysis in human beings, if spasms occur with these symptoms; some excitable parts behind the thalami optici must have likewise undergone some change.

To cure epileptic attacks caused by anæmia there is no better method than that of renewing the supply of red blood. The debilitating mode of treating epilepsy should almost always be rejected.

Circumscribed anatomical relations of the brain, or alterations of protracted duration, cannot be regarded as the proximate cause of epileptic attacks, but may cause epileptic *affections* (dispose to epilepsy). Pathological anatomy cannot give any explanation as to the nature of epilepsy. Suddenly withheld nutrition is only one of the causes by which the brain is brought into that peculiar internal condition which is manifested in the form of an epileptic attack."

They go on to say that while arterial congestion is unable to induce epilepsy, the symptoms of venous congestion are those rather of apoplexy than epilepsy. They hold that sphagiasmus and trachelismus are not the source of epileptic attacks, but that laryngismus may induce them, and that all theories are false which attribute the epileptic attack to a sudden determination of blood. They admit that spasm of the cerebral arteries may cause some forms of epilepsy, that the epileptic affection which disposes to the attacks occupies the whole brain, or some parts only; and also, lastly, that the medulla oblongata, as the part from which the nerves causing the constriction of the glottis and the vasomotor nerves take their rise, frequently seems to be the spot from which eclamptic and epileptic attacks proceed.

I must leave the opinions and statements of the distinguished physiologists to whom I have referred in the preceding pages, to the judgment of the scientific world at large. It is not for me to do more than thank them for the light they have, each in his way, shed upon the labyrinth in which so many have lost themselves. But I cannot but dwell upon the fact, that they all agree with regard to the essential identity of the various forms of epilepsy, which have been distinguished from one another according to the nature of the exciting cause. Whether the tubercula quadrigemina, the pons and crura cerebri, or the medulla oblongata, be regarded as the main seat of the disease, the irritation conveyed to them,—through the mind, from any other part of the nervous centres, from another viscus, or from a cutaneous nerve,—equally excites the paroxyms through the agency of the same part of the cerebrum. The real difference between different forms of epilepsy consists in the different excitability of the individuals: in one it is such that the perverted action once having been set up, it exhibits a self-multiplying power which cannot be controlled; in another, the excitability is so slight that the removal of the particular stimulus puts an end to all display of that excitability; while again, in a third, the excitability is moderate, but persistent, and liable to be called into action by any tolerably strong stimulus that may at any time be offered. It has been said, and I think with truth, that the possibility of producing epilepsy in every individual only depends upon the possibility of discovering the particular kind of stimulus which shall in him rouse his excitability to a certain point.

Given, the peculiar sensitiveness of the cephalic centre, which, on the application of a certain irritant, induces the paroxysm, what are the channels through which the irritant operates? Certainly in most, if not in all cases, it acts through the blood. A mere change in the balance of the circulation suffices in many instances; and if to that change a change in chemical constitution be added, the complication renders the physician's duty the more difficult. I might quote many cases illustrative of both these aspects. The following brief sketch of one may serve to show how epilepsy may depend upon an alteration in the balance of the circulation.

A young lady consulted me, through the mediation of Mr. Spencer Wells, in whom I could not but regard the disturbance in the balance of the circulation as the main element in the production of the epilepsy to which she had been subject for three years and a half. She is now (June, 1857) sixteen years of age, and has always enjoyed admirable health, with the exception of the fits, which possess the pathognomonic features of epilepsy. After a minute and searching inquiry, the only fact of any importance discoverable in connexion with the disease is that, although an older sister had menstruated at thirteen, she herself has not as yet manifested any symptoms of the catamenial function; and at the time at which she would have menstruated had she followed her sister's example, but at which in

P

her the fits made their appearance, the occasional attacks of epistaxis ceased to which she had been previously liable. One could scarcely avoid, in this instance, arriving at the conclusion that an *error loci* afforded the therapeutic indication; and that the disturbance in the circulation, acting upon a susceptible nervous system, deranged the polarity of the latter. As yet no material derangement of the nervous and intellectual functions are manifested; and were, from other causes, death to occur suddenly at this time between the paroxysms, I doubt whether it would be possible to detect any kind of lesion indicating an altered nutrition of the cephalic centre.

I believe that, in the great majority of instances of epilepsy, the first attack is due to an irritation produced by derangement in the amount or quality of the blood circulating in the brain. In a person predisposed we frequently find over-fatigue, a long walk, carrying heavy loads, prolonged mental exertion, the manifest cause not only of the first, but of many succeeding seizures. Hence there will be occasion, in discussing the treatment of the disease, to dwell much upon the necessity of bodily and mental rest, so as to allow the system to recover that balance, the disturbance of which gave rise to the seizure. This is more marked in some cases than others; but in none can our remedial agents be attended with any beneficial result unless we have a regard to this important indication.

These remarks lead to what I consider as of no less importance in the repetition and perpetuation of the disease, either from the point of view of its pathological theory or its therapeutic estimation—viz., the influence of habit. Medical men, no less than other observers of mental and physical functions, know how

THE INFLUENCE OF HABIT. 211

to appreciate this element in the influences to which mankind are subjected. For good or for bad, the repeated occurrence of the same acts facilitates their recurrence; and the proverbial expression, implying that force of habit may acquire an uncontrollable influence, is no less applicable to disease than to morals. It is a fact familiar to every medical man, that an individual part, which has once manifested a peculiar susceptibility, is prone to take on diseased action again; and that a frequent occurrence of disease in a part renders treatment more difficult at each succeeding attack. Hence the extreme importance of always thoroughly eradicating a morbid tendency when it first shows itself; and the no less important corollary which renders the period of convalescence from any disease, if possible, even more worthy of the special supervision of the physician than the stages of active disease. The reason of this is, that during convalescence the organs previously diseased are more liable to take on morbid action than other parts; and a return of disease in the former will not only leave the physician impaired forces to deal with, but, by the law of habit, will create an additional momentum in favour of the disease, which was before in abeyance. In no branch of pathology do we meet with more palpable instances illustrative of these observations than in the range of diseases of the nervous system. Habit is commonly interpreted as denoting a voluntary act which, by repetition, acquires the character of an instinctive act, though never entirely removed from the control of the individual if he chooses to exercise it. There is, however, a habit which, though never immediately under the control of the individual, may be subdued and held in check indirectly by the physical and physiological influences which we are capable of exercising. Such habit is the habit of an organ or organs to put on certain forms of diseased action.

In a disease like epilepsy, habit plays an undoubted and very important part. Every successive attack strengthens the habit, and renders the individual more obnoxious to future seizures; every arrest or postponement of a seizure is so much gain in favour of the patient, not only by avoiding the pain and the risk of the isolated paroxysms, but still more by diminishing his future liability to the disease.

Believing, as I do, that wherever we meet with epilepsy there is the same fundamental weakness of the cephalic nervous centre, and that, by repetition of the attack, the same ultimate results may be brought about, whatever the exciting cause may have been, the necessity of seeking by every means in our power to weaken, if we cannot succeed in breaking, the strong links which constitute habit, becomes an imperative law for the physician.

The case of the clergyman's son made a special impression upon my mind in reference to this point. The lad became suddenly and violently epileptic during the convalescence from scarlet fever, owing, as it appeared, not to any residuary affection, but to a trifling indiscretion in diet. Various evacuant remedies had been tried previous to my seeing him, to get rid of what was supposed to be the exciting cause. The attacks had continued, with scarcely any intermissions, for many hours, and there appeared to be imminent danger. The seizures were arrested by what might have been regarded as a hazardous remedy, by morphia. After the exhibition of the dose they subsided speedily, and they had not returned for eleven months after; and, had they occurred since, at a later period, I should certainly have been informed. Here was a case of hereditary epilepsy, for the father had also been subject to the affection; the convulsions were of the severest kind, and they supervened on a debilitating disease. This was assuredly a case in which an unfavourable prognosis was justified; and yet, by preventing the formation of a habit, the disease, terrible as it was, did not reappear. In speaking of treatment I shall again advert to the question of habit, as it is a point never to be lost sight of even in inveterate cases; yet I thought that my observations on the theory of the disease would have been even more imperfect than they are, had I not sought to dwell forcibly upon a feature which, under all circumstances, is one of vital importance.

In considering the theory of epilepsy, its relation to other diseases, and especially to those of a spasmodic character, must be borne in mind. There is much and powerful evidence to show that epilepsy belongs to a group of affections which are closely allied to one another, and hence exhibit many transition forms which have given rise to confusion in the minds of medical men.

The eclampsia of early childhood, laryngismus, or spasm of the glottis, may be especially mentioned as belonging to the same category as epilepsy. The main reason why, in infants, the convulsive character is not so prominent as in children of a larger growth, would seem due to that very impressionability which gives rise to the nervous symptoms on a comparatively slight stimulus. Their muscles and their spinal nerves have not reached that period of robust development which maintains later; while the slightest interference with the organs of respiration, dependent, as in the cases adverted to, upon spasms in the superficial or deeper muscles of the neck, causes loss of consciousness. We constantly see the gradations from the merest crowing inspiration to the most confirmed convulsive seizure in the same infant; while the recurrence of the well-marked epileptic seizure in the adolescent or adult is preceded in a sufficient number of times by infantile fits to justify the assumption of a close relation between the two. The following case may serve to illustrate the connexion between laryngismus and infantile convulsions.

The first child of a gentleman, at the age of six months, was weaned; ten days after, November 16, 1855, I was called to see her. For a week previously she had, both while asleep and awake, been seized frequently with an attack of crowing inspiration, accompanied by great anxiety and occasional lividity of the countenance; the attacks were very brief, and were followed by complete recovery. She had a slight seizure on waking from sleep during my presence, but without livor. There were a few mucous râles to be heard at the base of the left lung, the remains of a former catarrh. Some directions with regard to the regulation of the diet were given, an emetic of ipecacuanha, and the daily exhibition of steel wine, were ordered. The attacks at times occurred as often as twelve times in the day; but partly by the aid of medicine, change of farinaceous to milk diet, and country air, the glottic spasm appeared to subside. On the 26th December the account is,—the child was considerably improved, till two days ago she had a violent "epileptic" paroxysm; there was complete unconsciousness, the limbs were extended, the eyes turned up, and the thumb drawn in under the clenched fingers.

RELATION OF EPILEPSY AND OTHER DISEASES. 215.

The child did not scream. She became very pale, and the attack was followed by great prostration. Since this seizure the attacks of glottic spasm had been very frequent. Saw the child asleep; breathing quietly; head not hot; occiput rather warmer than vertex; fontanelle normal, not pulsating; motions hard and clayey; skin not hot. Habeat hydr. cum cret. gr. j., mag. carb. gr. ij., statim, and a wet-nurse.

Dec. 28. The motions yesterday were white, and more clayey than ever; in fact, the greater part of two was as white as snow. The fits continued. She had one severe one after the nurse came yesterday, at 8 P. M. The child took the breast with many objections, but had a good suck; and the first motion after was of a bright yellow, and feculent. Breathing calm; skin cool, and not hot. The "epilepsy" did not return, the crowing fits became gradually more sparing, and on the 4th January it is reported that she had passed twenty-four hours without any attack; on the 7th all the functions were healthy, and there had been no return. The child progressed favourably; under the care of an excellent wet-nurse she throve, till she was eleven months old. She then cut two incisors; a little catching in the breath occurred at this time, which would not have been noticed but for the previous attacks. While cutting the two upper incisors she, without being previously indisposed, had three severe epileptic attacks, at 2 A.M, 6 A.M, and 8 A.M, on April 8th. She was violently convulsed; there were carpopedal contractions; the eyes staring, the pupils dilated; complete unconsciousness, lasting several minutes; the fits preceded by a *gurgling* noise. There was no crowing. During the last fit there was considerable blueing of the features. The child was much heated; the skin dry; the bowels had acted, but the fæces were slimy, though bilious. The upper gums were hot and swollen, and on being lanced bled freely, the skin at once became cooler and moister. Some additional food had been given besides the breast, and was ordered to be omitted. Some castor-oil was administered. Four more fits occurred up to 5 P.M. on the same day: she then had a hot head; the motions were ill-digested, and the fontanelle was large and somewhat prominent. The hair was ordered to be cut short; ice to be applied; and hydr. cum cret. gr. j. tertiâ quâque horâ. April 9, half-past 8 A.M.; slept quietly the greater part of the night; there was no return of fits; the motions less slimy; is cheerful, and recognises those about her with a smile. The convulsions did not return, but the inspiratory spasm again supervened and continued, although generally improving in health, till the exhibition of Allarton's* steel-biscuits, which she took readily and regularly. From April 24 to June 3 there was no spasm of any kind; at the last date there was a sudden fit; a calomel purge, ice to the head, and an alkaline diuretic were the remedies employed, and she did well. The report on July 20 is: Has recently had several returns of glottic spasms; relieved by opiates and warm baths; the only traceable exciting cause was a slight bronchitic cough. She subsequently took cod oil, spent some time at the seaside, and when seen, Sept. 16, 1856, she is stated to

---

* These biscuits, which are manufactured by Mr. Allarton, 254, High-street, Southwark, are an admirable vehicle for steel, which is entirely covered by the pleasant flavour of the material. Each biscuit contains one grain of the ferrum reductum or Quevenne's iron.

have been in excellent health since the last previous report, July 30, which was also favourable; she could now walk well, and the fontanelle was closing. She continued the use of steel biscuits, and had daily baths, with the Kreuznach bittern. In November, after some premonitory restlessness, the child had two short fits; no irregularity of diet, undue excitement, or irritation of any kind could be traced. She had four double-teeth, and seemed to be cutting her right eye-tooth. The bowels were rather costive. After a purge and a diuretic, Boudault's *poudre nutrimentive* (pepsin, gr. v. bis die) was ordered, which she continued for a considerable time; it had a decided effect upon her digestion, and when the motions were out of order and contained undigested matters, they speedily recovered their healthy appearance under the use of the powder. From that time to the present, now about five years, the child has enjoyed perfect health, and there has been no kind of spasmodic action.

This case may serve as a proof of the intimate pathological relation between the ordinary convulsive fits of children and the glottic spasm, to which they are so often liable. Some of the severer fits in the case related were as much epileptic as any that I have seen in adults. It was also interesting as bearing upon the doctrine of the centric or eccentric character of the disease; because, while at times the brain was palpably affected, at others an eccentric causation seemed most apparent. The variations are Protean, but it is very important to appreciate the connecting links between allied forms of disease, as the success of our practice will be materially influenced by the view we take of the *rationale*. I have seen in a child, which subsequently died of glottic spasm, for a considerable time previously swelling of the left thumb, which was partially flexed across the palm, and which, as Marshall Hall pointed out in his Croonian lectures at the College of Physicians, is to be regarded as a symptom of the convulsive tendency. Such trifling manifestations of the disease may be easily overlooked; but whatever may be the prevailing view with regard to the epilepsy of adolescents and adults, nobody denies the curability of infantile fits, and the facility of recognising their approach necessarily increases our prospects of a satisfactory treatment.

Those who have watched the varying phenomena of neuroses with care will often have observed cases of metastasis, which are no less surprising than instructive. I speak not merely of the pains occurring in hysterical females, that pass rapidly from one part to another, but of definite affections of one class or set of nerves subsiding on the appearance of a similar or different affection in the range of another set of nerves. A short time ago I attended a lady, previously subject to intense pharyngeal neuralgia, for an intercostal neuralgia, simulating pleurisy. The former malady had entirely disappeared on the occurrence of the latter. In the course of my career I have seen many cases which appeared to me to support the same view, and the above cases may bear an interpretation of the same kind. We know—as Dr. Salter has more particularly shown—that centric asthma and epilepsy may be interchangeable diseases. He relates a very interesting case,* in which a man, aged fifty, who was subject to epilepsy, was on several

---

* On Asthma, p. 44. While these sheets are passing through the press a closely analogous case, occurring in an elderly lady, has been under my care.

occasions, when the premonitory symptoms of a seizure had presented themselves, attacked, not with an epileptic, but with an asthmatic paroxysm. Such instances also give support to the view of the relation of the respiratory tract in the medulla oblongata to epilepsy.

But there is another form of metastasis which we meet with every now and then, and which is not without a practical bearing both upon our prognosis and our treatment of the disease. I allude to those not unfrequent instances in which a disease of a totally different character, not a neurosis, but a genuine blood-disease—a fever, an exanthema—makes its appearance, being followed, not by a temporary arrest, but by a permanent cure of the epilepsy. These are inverse analogues of those cases in which eruptive and other affections are followed by chronic epilepsy. I have the notes of one very severe case of epilepsy, of many years' standing, causing incipient fatuity, which has been, to all appearance, permanently cured by an attack of measles, which proved all but fatal. I have before me an extract from the *Southern Medical and Surgical Journal*, (U. S.), in which Dr. Reagan gives an account of a young lady, who having been subject to epilepsy for several years, was attacked with remittent fever, in which she was severely salivated; the epilepsy was arrested, and had not returned three years after.

More such instances are recorded in the periodical literature of medicine, and deserve our special attention.

However we may wish to localize the affection, we cannot overlook the numerous facts that show the state of the blood to exercise a material influence in the production of epilepsy. We have had occasion to observe in a former chapter, that no uniform derangement is met with in the secretions, as far as we are at present informed, and which might serve as an indication of the special lesion which prevails. The close alliance between epilepsy and scrofulous affections points in this direction; while, in the great majority of cases, circumstances have preceded the outbreak which notoriously tend to impoverish the blood, and exhaust both the vascular and nervous power. But, although the disturbed polarity which induces the paroxysm most frequently depends upon exhaustive conditions—so much so that some writers, among whom I would specially mention Dr. Radcliffe,* regard this class of causes as the sole indication for treatment—I am satisfied that the state of the blood need not necessarily be impoverished, but that various pathological conditions of the blood may be associated with epilepsy.

Upon no other view could we understand the very successful result of the treatment pursued by Dr. Cooke,† and which consisted mainly in the abstraction of blood, not only on the derivative plan, but on the ground of diminishing the actual amount of blood in the system. His cases are well told; and from the description of his patients, who all seem to have been florid, robust country people, the reader will probably admit that the treatment was rational: unless the author be accused of downright falsehood, the success was palpable.

Other writers might also be quoted in support of the

---

* Epilepsy and other Affections of the Nervous System which are marked by Tremor, Convulsion, or Spasm. By Charles Bland Radcliffe, M.D. London, 1854.

† A Treatise on Nervous Diseases. By John Cooke, M.D., F.R.S. 1823.

view that epilepsy at times occurs under circumstances and in individuals justifying venæsection. This is certainly not my own experience; but the low type that appears to prevail generally at the present time in all diseases, and especially in the town population, among whom I have chiefly practised, may sufficiently account for this. Dr. Cooke formularizes his views thus: The predisposition to epilepsy, in whatever it consists, is evidently increased by a plethoric state of the body; though he justly observes, that the fact of the brain being found gorged with blood after the paroxysm does not prove the vessels to have been overloaded before. I shall revert to Dr. Cooke's cases again by way of illustration, and because they may tend to confirm what is so strenuously combated in other quarters, that diseases present a different type at different times and places.

In reference to epilepsy, both the quantity and quality of the blood must be taken into consideration. Many authors, among whom I would especially mention Dr. Prichard, support the doctrine that the immediate cause of the fit depends upon a preternatural influx of blood into the head. No instances are more striking evidences of this than where, as after hooping-cough, the fit is manifestly brought on by a retarded return of venous blood from the head. We have already seen that the prevailing post-mortem changes are such as cannot be explained, except on the assumption of a chronic form of inflammation, or at least of a change of nutrition generally connected with a larger amount of blood than average. The frequency of congestive headache, and the success of derivative treatment in many cases, may also be mentioned in evidence. Several cases are given by Prichard in which hooping-cough was the immediate forerunner of the seizure. However much we may be disposed to admit congestion to the brain as one of the accompaniments of the epileptic paroxysm, we cannot but see that, in the vast majority of cases of epilepsy, the patients are in a condition indicating a state of general anæmia, or a dyscrasic state of the circulating fluid—a state not certainly always, or even generally, to be measured by a physical standard, but no less recognisable by physiological tests—the state of the skin, the eye, the tongue, the pulse, the stomach, the intestines, the mental functions. In most acute diseases, even, that we have to deal with, we find that a predisposition is generated by previous debilitating influences—intemperance, debauchery, scrofulous or syphilitic taint, hereditary lithiasis. Such influences necessarily deserve equal attention in a chronic and periodic affection like epilepsy; and it is impossible to disconnect such influences, even in a disease so peculiarly in the domain of the nervous system, from a blood-lesion.

In concluding these remarks on the theory of epilepsy, I have pleasure in adverting to some observations by Dr. Handfield Jones bearing on the subject, which put it in a somewhat different light, though in the main they correspond with the views I have advocated. My friend's arguments will be better understood if taken in connexion with the case to which they are appended.*

" S. S., aged forty-five years, a plasterer, came under my care December 22, 1856, having been under

---

* Cases of Nerve-Disorder, recorded with reference to the probable Operation of Malaria as a Cause, p. 20. By C. Handfield Jones, M.B., F.R.S. London, 1856.

treatment for the previous ten days. He had been cupped at the back of the neck to eight ounces, had a seton put in the left arm, tartar-emetic ointment applied to the neck, and been purged, without benefit. He was of stout, rather short make; rather sanguineous aspect. He had been ill about four weeks, since a fall from a height of about eight feet. He states that he suffers from attacks of the following kind, occurring five or six times a day: He begins to wink, and then both eyes become drawn quite under the lids towards the right, one inwards and the other outwards; and he then loses consciousness for four or five minutes, and falls down: he does not scream, but moans as if choked. He has continual headache all along the right upper cranial region, from behind forwards. I witnessed the attacks more than once: they were perfectly involuntary; not attended by much flushing of head. He feels tremulous and nervous. He does not sleep well: has a trembling in the head when he lies down. Not an intemperate man: never drank spirits. Urine high-coloured; bowels open; pulse of good force; skin warm. He was similarly affected two years ago, and got well after six months. He had rheumatism twelve or fourteen years ago. He has not had syphilis, but gonorrhœa long ago. Tongue clean; eats food well. I gave him at first valerian with ammonia and iodide of potassium, and camphor with extract of St. Ignatius' bean, thrice daily. The camphor, &c., were soon changed for assafœtida, which was given at first in five-grain doses four times a day, and on Jan. 2 in ten-grain doses every two hours, and continued at the same rate till February 20, when he ceased to attend, having had only slight attacks in the previous fifteen days, and having been nearly free during several former weeks. The valerian and ammonia were given up when the assafœtida was given every two hours, but from January 9 to February 5 he took the following mixture:—

> ℞ Quinæ disulph. gr. iij.
> Tinct. ferri sesquichl. ♏xxv.
> Acid. hydrochl. ♏ij.
> Aquæ cassiæ ʒj.
> M. Fiat haustus, ter die sumendus.

He took this with an intermission of four days only. For the last fortnight, a mixture containing gallic acid, sulphate of zinc, infusion of calumba, and tincture of hyoscyamus, was substituted for the iron and quinine. He relapsed to some extent at one time when a severe frost gave way to a thaw. Remarks: I have cited this case partly because it is of much interest in itself, as a history of a curious neurosis. . . . . . In this case there was nothing of the kind (aguish disorder), and I do not feel that I have any sufficient reason for ascribing the phenomena to the action of malaria. The paroxysms evidently had much of the epileptic character, and the movements of the lids and eyeballs may be regarded as a kind of aura. The beneficial action of assafœtida, which seems to act as a peculiar cerebral nervine or toner, confirms very decidedly the opinion expressed by Dr. Todd, that 'the phenomena of the epileptic fit depend upon a disturbed state of the nervous force in certain parts of the brain—a morbidly-excited polarity.' What is wanted therapeutically is to find some drug or drugs which shall so influence the nutrition of the dynamic grey matter in the part affected, that it shall act with more steadiness, and be less mobile or excitable. How difficult this is, all experience declares. One great cause of this difficulty

I believe is, that brain-tissue partakes much more of individual peculiarity, and so differs much more in its vital endowments and in its reaction towards remedies than nerve-tissue does. We feel a great deal more confidence in our ability to cure a neuralgia than an epilepsy; still we trace much of the same features in the one as in the other; and our therapeutic proceedings are in principle the same. If excitement, excesses, bleeding, and all debilitating influences aggravate an epilepsy, so in most cases do they a neuralgia. The curative means in both cases are such as give tone without stimulating. But it may well be conceived that the tissue of the cerebral convolutions, whose function is of so high an order, and whose nutritive actions are so liable to be deranged and impaired by mental or emotional influences, is far less likely to be in all cases modified alike by the same agent than other tissues of lower and simpler functions."

The only other question which I think it necessary to open here refers to the propriety of establishing a distinction between acute and chronic epilepsy. Those readers who have favoured me with their attention during the preceding pages, will have gathered that I have represented the epileptic disposition as essentially chronic. And so it undoubtedly is, and yet we every now and then meet with cases which run so rapid a course, either to a fatal termination or to an apparently permanent cure, that we should fail to treat epilepsy in a manner analogous to other diseases if we did not admit that such a distinction does exist. And I here disclaim altogether mixing up cases of convulsions from uræmia and the like with epilepsy, but I speak of an affection presenting all the pathognomonic characters that we meet with in genuine epilepsy, except the one feature of chronicity.* I have already (at page 111) briefly alluded to a case that I should call acute epilepsy, which ended in recovery. The following is one that had a different termination; it occurred in the practice of one of my colleagues, and is given in the words of Mr. Myers, the clinical clerk :—

E. C., æt. 24, a servant, was admitted into St. Mary's Hospital on July 27th at 8 P.M. Her previous health had always been good; she presented the appearance of a robust person of plethoric habit. Six weeks ago she began to complain of pain in her head, increasing in severity, and towards night also of pain in her chest. After a time she became very sleepy during the day; and latterly, directly she sat down she would drop asleep. The appetite had always kept good; the catamenia had always been regular and proper in quantity. She remained in much the same state till the day before admission; she then had a violent trembling throughout the body which lasted about a quarter of an hour. At half-past three she had her first fit. In the following two hours and a half she had fifteen fits in succession, and did not regain consciousness in the intervals. Previously she had always enjoyed good health, excepting that just before her present illness she had scabies. After admission at

---

* Although there is no difficulty in diagnosing epilepsy from convulsions originating in diseased kidneys, if the attention is directed to these organs, it has happened within my experience that epileptiform seizures were regarded and treated as genuine epilepsy until the urine was examined. Serious mischief may arise from such an error. We should, therefore, never avoid instituting the necessary examination of the renal secretion in this as in all other morbid conditions. Too many chronic diseases especially fail to be correctly appreciated when this is not done.

8 P.M. the attacks continued to be very frequent until 12 P.M., when she became perfectly quiet and remained so till 8 next morning, when she became sufficiently conscious to recognise her sister, and call her by her name; in a few minutes she relapsed again into her former state, the fits becoming more frequent and violent, and they continued so throughout the day. In the intervals the breathing is quiet and the pupils contracted, but directly before the onset of the fit the pupils dilate to their full extent, and continue so during the attack, contracting again rapidly after it has passed off. During the fit her eyes are fixed; there are more or less violent general convulsions of the whole body and head, the fit generally commencing with a smacking and distortion of the lips; the tongue is bitten and there is froth at the mouth. The face gradually becomes intensely livid; after a minute or so deep inspiration takes place, frothy expectoration escapes from the mouth, there is great gurgling in the throat, and the patient then again relapses into the same comatose state as before. In the interval between each fit, which rarely exceeds five minutes, she is perfectly insensible; the pupils equal in size, contracted, and will not react to any stimulus. There is slight stiffness in the muscles of the arms when they are moved. Pulse 100, full. Skin hot and dry.' Heart sounds natural. Fæces and urine passed in bed. She is unable to swallow anything, therefore beef-tea injections are administered.

She was ordered calomel gr. viij. to be taken at once. Enema commune ʒxxx. with oleum terebinthinæ ʒj. Applicetur cataplasma sinapis cruribus et abdomini.

July 28, 2 P.M. A bleeding was ordered to 24 oz. Immediately previous to it the pulse was 152 and bounding; during the venesection it became softer and fell to 128; directly after, it again rose to 160, and became smaller and softer. The blood which was first drawn was very dark in colour, but the second portion was much brighter; it coagulated rapidly and gave off ammonia, as shown by testing it with hydrochloric acid. No change followed in the character of the fits after the bleeding. She was freely purged by the calomel. At 6 p.m. the pulse was 168 and soft; the fits still frequent, but not so violent.

July 29, 11 a.m. The fits continued throughout the night with but small intermissions until 8·15 this morning, since which time the patient has not had a complete fit. The attacks now consist more of convulsive spasms of the various muscles, sometimes affecting only those of the face, sometimes those of the extremities also. She does not now become livid, and there is now no rattling in the throat. The pulse is very irregular, at one moment beating rapidly, at the next very slowly. Respiration 67 per minute. She keeps her eyes open and rolls them about. The muscles of the arms become very stiff on attempting to move them, and the pupils act slightly; they do not dilate at the commencement of a paroxysm, and they are not so contracted in the intervals. She swallows fluids in small quantities, and appears to taste them. 8 p.m., the feet and hands have become very cold.

July 30. She sank at 2 a.m., no apparent change having taken place since yesterday morning. She died in a convulsion, and was quite blue in the face.

The autopsy was made by Mr. Gascoyen twenty-four hours after death.

The body was that of a very short woman, very broad and fat; complexion dark and foreign-looking.

Vesications over the abdomen from a mustard poultice; in the bend of the left elbow was a recent lancet wound; decomposition had commenced. Head: on removing the calvarium fluid escaped, and there were about 2 oz. of discoloured serum at the base of the skull. The vessels on the surface were congested, as were the sinuses; the brain itself, which was examined with great care by Mr. Gascoyen, was healthy throughout. On the left side there was a small hard cretaceous mass the size of a moderately large pea, on the base of the anterior lobe a little external to the optic nerve. Weight 3 lbs. 3 oz. The other organs were unfortunately not allowed to be examined.

Should any of my readers hesitate to regard the above case as one of epilepsy, I append the following brief summary of another case, seen in consultation, as an instance of the acute variety, concerning the epileptic character of which no doubt could be entertained:—

M. D., a gentleman, æt. 41, having no hereditary taint, but for some time past affected with chronic rheumatism, was in his usual state of health when, three days before I saw him, he was suddenly seized while riding in an omnibus with what, according to the description of a relative, was concluded to be an epileptic attack. He completely lost his consciousness, and on recovering himself was somewhat oblivious. After supper on the same day at 8·30 he had a slight attack of vertigo and nausea. The day after while in the street he had a severe seizure; he screamed and completely lost his consciousness, and continued for half an hour after recovery in a state of partial insensibility. In the evening after supper, consisting only of soup, he had nausea, followed by a slight attack. He slept well, but on waking up in the morning he had lost all memory of the events of the previous day. At 9 A.M. on the day before the consultation, he had a brief attack of unconsciousness; at 4·45 P.M. of the same day he had a severe seizure, with screaming, frothing at the mouth, and complete unconsciousness. Between the time of the first consultation on the 24th of March, and the second on March 26th he had five seizures, in one of which he was seen by Dr. Gueneau de Mussy, the physician in attendance, who stated that there was no doubt about their epileptic character. These attacks were described by the relative as being milder than the earlier ones. After the treatment, as sceptics would say, but in consequence of it as I believe, the seizures were now arrested, and did not return up to the 5th of May, and I believe that the gentleman has remained free from them since. I say nothing of any other symptoms of the case, or of the treatment now, except that it was directed against what we believed to be the source of irritation, a rheumatic meningeal affection of the spinal cord. I shall revert to the measures adopted in this particular case in the chapters on treatment.

Although I do not think that every variety of epilepsy is comprised in the above remarks, I am of opinion that they apply to the great majority of epileptic and epileptiform seizures that we have to deal with. Some of my remarks will be supplemented in the following chapters devoted to the treatment of the affection. The difficulties that we shall meet there are great and serious; but there are many earnest labourers, whose work I believe to be in the right direction, and to promise ultimate and complete success.

We must in the mean time bear in mind that, as Dr. Handfield Jones justly observes, the reaction of the complex tissues of the brain to different remedial agents varies much more in different individuals than the reaction exhibited by other organs and tissues. Hence the great diversity of remedies that have been lauded in the treatment of epilepsy, which certainly for the most part possess some analogy in their characters, but which may not unreasonably be supposed, under different circumstances and in different individuals, to have wrought different results. In this point of view, the apparent discrepancies of different observers may possibly be reconciled. I shall be unable in detail to examine even a tithe of the substances to be found in the *armamentarium epilepticum:* this would be impossible without much expanding the proposed limits of this work.

I do not, however, wish it to be inferred that, if I pass over some drugs that have been highly extolled, or say less of others that have appeared valuable in some hands, than they may deserve, I deny that cures have been effected by many of which I see the *rationale* even less than of those that I have most faith in.* Faith, however, is an abstraction which ought not to be our guide as to the action of medicines. But before science will substitute for it a positive knowledge of therapeutics, we must know more than

---

* Romberg (l. c. p. 228) states that above fifty years ago anti-epileptic remedies were so numerous as to fill 150 quarto pages in Hemmings' "Analecta Literaria Epilepsiam Spectantia." More accessible works are those of Tissot, Cooke, and Copland, which contain tolerably complete lists of the remedies that have been tried.

we do of the physiology of the nervous centres, and of the definite reactions which take place in them on the administration of definite physical agents.

In the meanwhile I offer, with a full consciousness of their imperfection, the following observations on the treatment of the disorder.

CHAPTER IX.

The treatment of epilepsy—During the paroxysm—Removal of all restraint—Avoid over-active treatment—Cool air—Sinapisms and the like—Compression of carotids—Cold applications—Galvanism—Volatile stimulants—Treatment of premonitory symptoms—Illustrative cases by the author—Ligatures—Dry-cupping—Internal remedies—Radical treatment of the disease—Trephining—Moral treatment—Derivation from the head—Various counter-irritants: the actual cautery, blisters, setons, tartrate of antimony ointment—The abstraction of blood—Illustrative cases—Ligature of carotids—Purgatives—Turpentine—Salts of iron—Cases—Zinc—Cases—The valerianates—Silver—Dr. Frommann's case—Nitro-muriatic acid—Digitalis—Iodides—Bromides—Cases—Belladonna—Indigo—Cotyledon umbilicus—Mistletoe.

IN approaching the complex subject of treatment I would make one preliminary remark. It is that the medical man should from the moment he has clearly recognised the malady, whether it appears in its most frightful form, or whether it presents itself in the guise of a comparatively slight attack, not dally with it; nor let him, to save the feelings of the relatives from the shock which they will feel all the more if precious time is lost and they are apprised of the nature of the disease at a later period, conceal from them its real character. Incalculable mischief is thus often wrought, and time is lost that cannot be recovered.

In discussing the therapeutics of the malady, my

endeavour will be to express myself as clearly and as concisely as the subject demands.

The treatment of epilepsy resolves itself into two main considerations—the proceedings to be adopted during the fit, and the course to be pursued in the intervals.

The main object in the paroxysm is to prevent the patient from being injured by the violence of the jactitation. He should be placed on a couch, with the head raised; all mechanical restraint ought to be removed which may interfere with the circulation, especially all ligatures and confining media; neckerchiefs, stocks, cravats, stays, should be loosened. As epileptic patients are apt to injure themselves severely in their fall, they ought never to be left alone, especially not in a room with an open grate, as it but too frequently happens that the paroxysm has caused the patient to fall into the fire, and to be grievously burnt. A fire-guard is even more necessary in the rooms occupied by epileptic patients than by others. Tissot relates a curious case of a child whom he saw, and who became epileptic at the age of eighteen months, in consequence of a pistol being fired off close to its ear. At the age of three years the child was cured by accidentally sitting down upon a brazier of live coals, and burning his posterior. The author remarks that this is the only case in which a burn has cured a patient. I have met with no analogous case either in my own practice or in that of other writers, unless we regard it as an illustration of the value of counter-irritation by heat.

Epileptics fall down stairs, into areas, into the water, or from horseback, when their fits supervene without warning. Many cases of drowning are attributable to epileptic seizures coming on while the patient is in the water. Hence no epileptic ought to be allowed to swim, and even the ordinary hip-bath may prove dangerous if the patient, being attacked, fall forward with his face into even a few inches of water. Clearly those patients cause less anxiety who are fortunate enough to experience a warning in sufficient time to enable them to place themselves so as to be out of danger.

Having secured the patient in a safe position, untied neckerchiefs and loosened waistbands or stays, and provided for a copious supply of cool, fresh air, we may apply cold water or cooling lotions to the head, but should avoid all restraint that may not be necessary to prevent the patient from doing injury to himself.

If there are recurrent fits with symptoms of much venous congestion, powerful counter-irritation in the form of terebinthinate fomentations or sinapisms to the extremities, and leeches to the temples, are to be recommended. The application of ammonia and other stimulants to the nostrils is to be deprecated; and it is always to be borne in mind that the unconsciousness of our patients prevents their giving timely warning of injury that may be inflicted upon them. A young gentleman under my care recently suffered for a considerable time from a burnt foot, owing to the attendant having applied hot bottles to the feet, and having forgotten to protect them by flannels.

The spasm which forces the tongue out of the mouth, and then firmly closes the jaws upon it, often causes great anxiety lest the member should be bitten off. The attendant must carefully watch the tongue, and, if it is protruded, push it back without delay, or watch his opportunity, and introduce a cork, or piece of wood or india-rubber into the mouth, so as to serve as a gag.

When the spasm is severe, I do not hesitate to employ chloroform by inhalation; it relaxes the spasm, and when this effect is attained, its use should at once be discontinued.

The temptation to employ active treatment is necessarily great; and it is painful to stand and watch the epileptic paroxysm without being able to suggest an efficient remedy. Perhaps more might be done if physicians more frequently saw their patients during the attack itself; but the suddenness of the seizures, and their generally brief duration, render it impossible in most cases to apprise the medical attendant of the occurrence until it has passed by.

The compression of the carotids in the epileptic paroxysm is a proceeding of very doubtful propriety, especially if we hold that in epilepsy we always want more rather than less arterial blood. In one of the cases in which I have used it experimentally, the attempt to compress one carotid was immediately followed by a severer fit than the one that preceded. Still, as with chloroform, which is strongly disapproved of by one or two writers, it is a matter of empiricism rather than theory.

Dr. Caleb Hillier Parry* suggested and frequently employed compression of the carotids in head affections, with the most beneficial results.† He states that he has often, by pressure on one carotid, arrested tinnitus aurium on the corresponding side of the head; and that when any degree of pain has existed about the

---

* Collections from the Unpublished Medical Writings of the late C. H. Parry, M.D., vol. i. p. 392. London, 1825.

† Dr. Fleming has recently drawn attention to the effect produced by compression of the carotids, *British and Foreign Medico-Chirurgical Review*, vol. xv. p. 529.

neck and throat, strong pressure of the carotid on one side has always increased the pain on the other, owing to a larger quantity of blood passing through its fellow. "When there has been a sense of weight, and fulness of the head, and flushing heat in the face," he says "I have many times felt all taken away by pressure on one or both carotids; and the feet, which were before cold, by this change of determination, experienced a comfortable glow of heat." Dr. Parry has seen epistaxis moderated, and sleeplessness and excitement removed, by this compression; but he does not, although he makes some valuable remarks on the subject of epilepsy, appear to have employed it in that affection.

Dr. Romberg has found compression of the carotids an effectual prophylactic in patients who have forewarnings, and are able to employ it in time. One of his pupils even invented an instrument for the purpose of compressing the carotids, but it failed in its object. The only case besides those mentioned by Romberg that I have been able to discover, in which this proceeding was employed in epilepsy, is one stated by Dr. Prichard to have occurred in Mr. Earle's practice, where a cure was effected by bleeding and purging, but where temporary benefit was obtained in the paroxysms by compressing the carotids. As an arrest of the return of blood from the brain is a manifest concomitant of epilepsy, it may be urged that this proceeding may prove useful in epilepsy, by diminishing temporarily the supply; well-trained attendants might be permitted to employ the method at least on one side of the neck; but it would scarcely be right to permit its use to any casual attendant of the invalid. The proceeding requires further corroboration to justify its general employment. Should there be a class of cases in which it proves beneficial by arresting or abridging the fit, they should be clearly defined; or if it is of advantage where a premonition gives time for the application of a remedy, it would be no less acceptable. The postponement of a fit, and every diminution of the severity of a paroxysm, is a gain; and if Dr. Parry's observations are confirmed, it is probable that the cephalalgia and somnolency, which the patients so frequently complain of as distressing symptoms following the attacks, might be thus diminished.

The occurrence of the fits during the night may, if not controlled, be occasionally held in abeyance by the application of cold lotions to the head; and wherever we find the head hot, this indication is still further followed out by clipping the hair quite close. I have before had occasion to allude to a case which fell under my own observation, in which it appeared that a dose of morphia arrested a most violent series of paroxysms; it might have been a coincidence, but whether or not, it is a fact worthy to be remembered, since, as the patient certainly was not injured, the agent might be applied again under similar circumstances. As a preventive, narcotics and sedatives certainly have failed to exercise any beneficial influence in my hands; and they are less spoken of by writers on epilepsy than probably any other class of remedial agents.*

While deprecating the nimia diligentia medici (vel amicorum), I am satisfied that the appliances above spoken of occasionally do real good. Those medical men wrong their science and their patients who encourage the public to neglect all means of shortening the paroxysm; for the exhaustion and consequent in-

---

* Dr. Matthew Corner has recently advocated the use of opium in epilepsy. *Lancet*, April 16, 1859.

jurious effect to the whole body, and the nervous system especially, are in the direct ratio of the duration of the fit. From my personal experience of the use of the galvanic current in muscular spasm as a tonic and sedative, I should expect it to be useful during the fit. The relief it affords in chorea is indisputable; and in epilepsy I have frequently administered it to combat the muscular pain and exhaustion left by the paroxysm, to the great satisfaction of the patient. During the paroxysm itself I have not yet applied this agent, but purpose doing so. In this case the conductors should be moistened sponges, so as to insure the passage of the current to the deeper-seated tissues. If we apply the metallic brush, we should cause irritation on the surface, and by stimulating reflex action, probably augment the paroxysms.

My ground for making this remark is that I have, on several occasions, found the application of the metallic brush to the dry skin excite symptoms of an approaching paroxysm in persons who exhibited a distinct aura. In a young married woman subject to severe epilepsy for some years, and in whom there was a well-marked premonition passing upwards from the right foot, the application of the galvanic current to this extremity speedily produced the aura, with all the sensations of an approaching fit. In a young gentleman in whom the aura proceeded from the left hand, and extended up the corresponding arm, the same effect resulted, on making the application to the part. In a gentleman in whom the fits were preceded by spasmodic action of the various limbs promiscuously, the application of galvanism also produced sensations like those that characterize the approach of a fit.

Little as we can do of a positive character during the fits, that little often requires regulation. And especially it is necessary to protest against the vulgar notion that by forcing open the clenched fist, or overcoming any other spasm of the body, by sheer violence, we may shorten the paroxysm. The notion still prevails, and should be combated by medical men.

After the fit has exhausted itself, nature commonly seeks restoration in sleep, which need not be interfered with unless symptoms of injury to the brain are manifested. These would be treated according to their nature and severity, and demand no especial notice. Injuries to the tongue or lips heal rapidly without interference, and other lesions inflicted during the seizure demand those surgical appliances which are indicated according to their severity. In the majority of cases, rest of body and mind are all the precautions that need be enjoined; but if the exhaustion is great, some volatile stimulant, or a more persistent tonic, may at once be necessary. I have found galvanism applied along the spine with moist sponges afford comfort to numerous patients at this time in the most undeniable manner.

An interesting question is involved in the treatment of the premonitory symptoms of epilepsy. It has long been held that, when the *aura* can be arrested before it reaches the brain, the fit will be prevented. And, whatever may be the theory, or however we may wish to escape the fact, the evidence appears undoubted. We have seen that premonitory symptoms do not always occur, the proportion being about equal of epilepsy with and epilepsy without premonitory symptoms. They may be divided into two kinds; those which have a distinctly local character, and those that consist in a more indefinite or indefinable general sensation. In both cases we may often succeed in preventing the supervention of a fit by employing appropriate remedies. Where the indication of an approaching fit assumes the character of a sensation passing up one of the extremities, means can be taken to arrest the phantom current. An amusing instance is given by Dr. Lysons* of a farmer's daughter to whom he was called, and in whom the fits always showed themselves by an aura proceeding from the feet, in consequence of which he ordered strong ligatures to be applied below each knee. "The method had the desired effect; the epilepsy proceeded no further than the ligature, but the feet shook most violently, and made so ridiculous an appearance, that the girl herself, though in the greatest distress, could not refrain from laughing heartily, and almost at the same instant begging us to let the disease take its course, lest her feet should drop off by the violence of their agitation, which, she said, was intolerable." The repeated use of the ligature in this case was followed by a complete cure. The result confirms the justice of our remarks upon the influence of habit in perpetuating the disease, and the propriety of our seeking for all legitimate means by which the habit may be interrupted. Romberg† quotes a similar and an instructive case from Odier, of a soldier " who, after having received a sabre-cut across the left side of his head, often suffered from spasmodic contraction of the little finger of the right hand, which subsequently extended to the fore-arm, shoulder, and neck, and each time ended in an epileptic seizure. After having tried numerous remedies without any result, Odier advised him to bind a cord

---

* Practical Essays upon Intermitting Fevers, &c. By Daniel Lysons, M.D., Physician at Bath. Bath, 1772.

† A Manual of the Nervous Diseases of Man, vol. ii. p. 212. Translated by E. H. Sieveking, M.D.

tightly round the arm at two places, between the elbow and wrist, and between the shoulder and elbow; by following this suggestion the epileptic paroxysms were staved off for three years." Unfortunately on one occasion, in a fit of drunkenness, the patient paid a severe penalty for his intemperance; for having forgotten the usual precaution, he died during the attack that followed. The post-mortem showed a severe intra-cranial lesion.

I have met with numerous instances illustrating the great utility of adopting similar measures. The following cases are given in full, in illustration of this point, because they also serve to mark the value of other methods of treatment to which a reference will be made later on.

E. W., æt. 24, the wife of a builder, came under my care for epilepsy Feb. 19, 1858, from which she suffered for four years. She had always enjoyed good health, the catamenia occurred at an early period, she had never had a blow on the head, but at the time of the first occurrence of the fits, about nine months before marriage, she was labouring under much anxiety owing to the illness of her parents and other domestic affliction. Previous treatment had been of no service. Since marriage she has had but one child, and during her pregnancy the fits have been less frequent than at other times, and they did not occur during her confinement. They generally take place once a month, but during the last fortnight she has had two. The fits most frequently seize her in bed; she is then awakened by a pain in the right toe, and this mounts up to the abdomen. Sometimes by getting up and rubbing the foot the sensation passes off. When the numbness of the foot comes on, the hands perspire very much. The premonitory symptoms last four or five minutes and more. During the fit the right foot is drawn up to the body, the left foot not being moved. The fit is accompanied by a scream, and the patient bites her tongue. If she stoops before the fit, when the premonitory symptoms come on, the fit follows more rapidly. There is complete insensibility during the attack, but before this supervenes she feels as if she were screwed up into the air; and then something seems to burst in her head and the unconsciousness is then complete. The catamenia are regular, or at times too frequent. There is trifling numbness or a sense of fluttering in the right leg, and pain across the renal region. The urine thick; the tongue furred; appetite fair; pulse 101, and very feeble. She was ordered to apply a strong cord round the leg so as to be able to tighten it instantly when an incipient aura was felt, and the following mixture.

℞ Olei terebinthinæ ʒiss.
Mucilaginis ʒij.
Aquæ pimentæ ad ʒxij.
Uncia ter die sumenda.

Feb. 23. Has had no fit; three days ago the ligature served to prevent one that was approaching. There is severe pain in the leg extending down to the toes. There is bearing down and dysuria from the mixture; the urine dark-coloured, and some pain in the region of the kidneys.

℞ Sulph. zinc gr. ij.
Ex. inf. valerian. ʒss.
Ter die sumendum.

Feb. 27. Urine examined; had a sp. gr. of 1027, being collected from the previous evening. It contained copious lithates, but no albumen. Treated with equal parts of nitric acid, a small quantity only of nitrate of urea formed after a lapse of time, smaller in amount than I have met with in other specimens of a lower sp. gr. It contained epithelium from various parts of the urino-genital tract, and some mucous corpuscles. There was some tenderness in the region of the kidneys, and in the lumbar vertebræ, but not in the sacrum.

March 2. Had a severe but short fit this morning; the use of the leg is almost gone, so that she has a difficulty in walking. Repeat mixture; add sulph. zinc. gr. ij.

 ℞ Ol. menthæ ♏vi.
  Ol. terebinth. ʒvj.
  Ol. ricini ℥j.
  Mist. mucilag. ad ʒvj.
  Tertia pars bis hebdomade sumenda.

March 5. Has had four sensations in her foot as of a threatening fit, but the bandage on the leg each time prevented it. Otherwise the same.

March 12. Has had the premonitory symptoms about five times, but each time anticipated the fit by the ligature. The ol. terebinth. disagrees.

 ℞ Oxidi. zinc. gr. vj.
  Acid. lact. ♏xxx.
  Inf. valer. ℥ss. ter die sum.
 ℞ Pil. coloc. co. gr. iv.
  Strychniæ gr. $\tfrac{1}{12}$
  Omni nocte cap.

March 16. Has had no fit, but the leg has been very painful to-day; there is a sense of great weight all over the extremity, especially at the calf and over the dorsum, but there is no visible or tangible lesion. In walking she feels as if she were going to fall. Is sick after the medicine whenever she takes it. (This intolerance of zinc is very rare in epilepsy).

> ℞ Zinc. oxid. gr. vj.
> Ferri carb. c. sacch. gr. viij.
> Sumat pulv. ter die.
> Rep. pil. omni nocte.

March 19. Has had no fit, but there is much pain in the leg, like a dreadful dead weight. Yesterday she had the ligature applied five or six times to prevent a fit. Rep. pulv. et pil.

March 30. No fit, but repeated premonitions, and much pain in leg. The powders sicken.

> Rep. pulv.
> ℞ Linim. terebinth. ℥ij.
> Ol. croton. ʒss.
> Fiat linim. regioni sciaticæ infricand.

April 6. There is severe eruption on the thigh, with but little relief to the pain. No fit, but the sickness after the powders continuing, they were changed for phosphate of zinc gr. vj. ter die. It was now above a month since she had a seizure.

April 27. She has been absent in the country, but has taken the powders regularly. She has had two fits during her sleep, and was therefore unable to apply the ligature. There is much pain in the sciatic, but it is relieved by the liniment. The last fit occurred five days ago, since which time she has been worse. She struck her back severely in the last fit. The powders cause vomiting and nausea.

℞ Atropiæ sulphatis gr. j.
Solve in spir. vini ♏D.
Cujus solutionis sume ♏xx.
Aquæ piment. ʒij.
S. Cochleare amplum omni nocte sumendum.
(Dose, one-hundredth of a grain.)

April 30. Has been very ill since last visit, having had many strange sensations, and yesterday a loss of power in her legs for an hour. The fits have been kept off by binding the leg as soon as the symptoms were perceived. She complains of her head and great depression, with severe pain in the thighs. Last night she had a curious and fearful sensation, such as she has not had before. The patient attributed her unusual depression to the atropine, which was omitted. Zinc was again ordered in five-grain doses, to be increased by two grains every fourth day.

I now lost sight of her till the end of last year, when she again came under my care, with the same symptoms, except that the fits had again become more frequent, occurring every two or three weeks. The premonitory symptoms continued the same, and she complained especially of pain over the right foot. It is unnecessary to repeat the details.

Nov. 2, 1860, I prescribed the following:—

℞ Bromid. potass. gr. x.
Extr. belladonnæ gr. ¼.
Aq. pimentæ ʒj.
Ter die sumendum.

℞ Pil. coloc. co. gr. v.
Alterna nocte.

Nov. 6. No attack, but much pain in right leg and foot.

Rep. mist. et appl. vesicat. 3 × 2 parti cruris dolenti.

To abridge the account I may state that a series of blisters was applied over the leg and foot, with great relief to the pain, the ligature being also employed above the point at which the pain commenced, when the fit threatened. The right leg, which had become smaller and weaker than the left, as ascertained by measurement, was next treated with frictions with linim. camph. co., which strengthened and manifestly improved its nutrition. The same mixture being used all the time, no fits occurred until the 7th of Feb., or about fifteen weeks, a considerable gain upon an interval of two or three. The fit which then occurred could scarcely be called a complete one, for she did not become unconscious, nor scream, or bite her tongue. The last report is dated March 26, 1861. There has been no return of fits since Feb. 7, although she has occasional painful sensations in the leg. Only two days ago she twice stopped the attacks by the ligature; but even when prevented she feels poorly for a few hours, though not as bad as when she has a regular attack. It is now over six weeks since the last complete seizure. She has been much better on the whole. Her general appearance is much improved. There is less wasting of the right leg, but there is a tender point at the inner side of the right leg at about its middle, close to the tibia, corresponding to the internal saphenous nerve; from this point the pain at times goes up towards the crural ring or sciatic notch. Two small blisters were again ordered to be applied to this spot, and the old bromide of potassium and belladonna mixture to be persevered with.

The next case is that of a medical officer of the army, aged forty-one, rather stout, and dark complexion, whom I first saw Feb. 17, 1859. He had formerly been in the West Indies, where he suffered from enlarged liver, and was sent home with albuminuria, from which he recovered. During the Russian war he was ill in the Crimea with fever, but has otherwise enjoyed fair health. There is no hereditary predisposition, and though he used to smoke much, he was temperate in other respects. In June, 1857, at Gibraltar, while hunting, he fell to the ground in an epileptic fit, and has since then had five attacks, which are generally attributable to severe bodily or mental fatigue. He has had tapeworm twice, but has seen no trace of it lately. In one fit he bit his tongue. There are no head symptoms, but at times cramps in the left extremities, especially in the thigh. During the fits, the left side of the face is said to be much convulsed. The pulse is of average strength and frequency, the heart normal, and there is no enlargement of the liver or spleen. He is much depressed on account of the disease, and its possible consequences to himself and his family. He was advised to apply the ligatures to the arm and leg, and to take twenty drops of the following three times a day in water :—

℞ Tr. ferri muriat. ʒvj.
Ether. chlor. ʒij. M.

Jan. 16, 1860. The patient continued well for seven months after the last visit, when he had an attack. During the following Christmas he suffered from severe tonsillitis, and before he recovered from this he was seized, without any premonitory symptoms, with a very violent attack, or rather series of attacks, on Jan. 3. The urine has a sp. gr. of 1012; it is pale, and contains copious oxalates. The pulse is eighty; there is flatulency, but generally a good appetite. He states that he has found much benefit from the ligatures which were ordered to be applied at the last visit, when the spasms and sensations affect the left arm and leg. They are more frequent in the arm than the leg. Belladonna with zinc and nitro-muriatic acid were now substituted for the former drops. A month later there had been no return of the attack; the pulse was reduced to sixty-eight, the appetite was good, and with the exception of one or two attacks of vertigo, he had nothing to complain of; the belladonna had been taken nightly in gr. ¼ doses. Since then I have heard nothing more of him.

The following is the history of an epileptic female whom I only saw once, but who stated that she had found application of a ligature to the extremity arrest the fits:—

A. K., a laundress, æt. 38; tall, well-built, and of healthy appearance, always enjoyed good health till she was about eighteen or nineteen years old. She then was attacked with epilepsy, supposed to have been brought on by tænia, which, however, was removed without removing the fits. The fits were at first severe, occurring every six to eight weeks; but every week there were attacks of spasm in the right hand, like incipient fits. They always commence with formication in the fingers, which mounts up to the hand and the arm. Before the fits came on she had been subject to this formication from the age of ten. This aura lasts about two minutes, when the fit supervenes with a scream; the head is drawn to the right side; there is complete unconsciousness for a quarter of an hour; the tongue is bitten, and she foams at the mouth. The fits are followed by headache and somno- lency, but there is no headache during the intervals.

Before the fits she has a sense of terror. During her pregnancies she is free from attacks, but they are worse when she is suckling. They occur mostly by day, rarely by night, and they have not become more frequent of late. At the time of the consultation she concluded herself to be pregnant, from the fits not having occurred for six weeks. I advised the application of a ligature to the arm, which she stated she had already used with great benefit, inasmuch as it arrested the aura, and prevented the fit.

Where the advent of a seizure is heralded by giddiness and symptoms of active congestion, which, *pace* Messrs. Kussmaul and Tenner, I am perfectly certain do occur in some cases of genuine epilepsy, the application of derivatives in various forms is of undoubted value. I know of none equal to dry cupping applied to the upper part of the back and to the nape of the neck. There is not often time in such cases to send for a medical man. I therefore advise the purchase of a little cupping apparatus,* with an exhausting syringe, which can be applied by the wife or anybody at all acquainted with it. This can be used at a moment's notice, and often gives great and instant relief. The following is an illustrative case:—

The Rev. S. T., a strong, robust-looking man, æt. 29, without hereditary taint, had always enjoyed very good health till the beginning of 1859. In August, 1858, he had been thrown out of a car with his wife, who was severely injured, and the illness

---

* Mr. Becker, of the firm of Elliot Brothers, the eminent philosophical instrument makers in the Strand, constructed such a one for me many years ago, and my patients find it very convenient.

which followed brought her to death's door, so as to cause him great anxiety. He was not himself injured on the occasion. The anxiety of watching over his wife induced vertiginous attacks, with flashings of light. In February, 1859, the first fit occurred, preceded by vertigo and flashes of light, followed by complete unconsciousness and convulsive struggling of both sides of the body equally; he first becomes pale and then livid, and snorts. There is no scream, foaming, or biting of the tongue. He does not recover himself for a considerable period after the attack has passed off. He was treated with marked benefit with counter-irritation by tartrate of antimony to the neck, and laxatives, with various preparations of steel, strychnine, and other remedies, and subsequently went into the country, and I have not heard from him for a considerable time. But in his case the use of the cupping glasses was very marked. They were applied to the nape of the neck; and having obtained a case for himself, he often arrested the vertigo, and as he himself thought, prevented the epileptic paroxysm by being dry cupped.

When the premonitory symptoms are entirely subjective, and depend upon some depressing effect produced upon the nervous system, whether it is merely a sense of fear, or flatulency, or some indescribable sensation appreciable only to the patient, stimulant drops and draughts are of decided use. Patients and their friends undoubtedly often attribute effects to wrong causes; but the frequency with which they have assured me of a fit having been staved off by the employment of such means justifies me in attributing some influence to remedies like sal volatile, chloric ether, valerian, castoreum, assafœtida, and the like, by themselves, or in various combinations, as, for instance, in the following prescriptions:—

℞ Spir. ammon. arom. ʒss.
Ether. chlor. ʒij.
Tr. castorei ʒiij.
Aquæ flor. aur. ʒvij.
S. cochleare parvum pro re nata ex aquâ sumendum.

or

℞ Tr. lavand. co.
Tr. castorei.
Spir. ammoniæ aromat.
Syrupi zinzib. āā ʒiv.
S. cochleare parvum pro re nata et aqua sumendum.

The will is sufficient in some patients to ward off an attack; in others, dashing cold water over the head and face has a revulsive effect.

Many cases are on record where local treatment of the kind above spoken of has prevented or cured the disease. The removal of a testicle, the amputation of limbs from which the aura seemed to proceed, has been followed by recovery. In some of the instances recorded there is proof of an irritation having been kept up at the part by disease or injury; in some it is probable that the part was as little the actual source of irritation as the arm in M. Odier's soldier. There is less difficulty in understanding the *rationale* of trepanning in some cases of epilepsy, in which, by injury to the skull, causing depression, or the formation of an exostosis, a source of local irritation of the brain is established. When we have discovered some means of determining the site of intracranial lesion with more certainty than at present, we may hope to reintroduce trepanning in the treatment of epilepsy, and to employ it more extensively than it is now used. If there is a fair probability of our meeting with the irritant, we may perform the operation, but we always run a risk of searching in vain. However, the Count Philip of Nassau-Weichem will always remain a proof that trepanning is not in itself an operation to be feared. He was not epileptic; but having suffered from symptoms of compression after a fall from his horse, without any indication of the exact spot at which the effusion had taken place, the trephine was applied at random; it was not until the *twenty-seventh* application that the site of injury was discovered. This fortunate Count, who lived in the middle of the seventeenth century, survived many years after the operation, and " could even drink more wine than before without being drunk."

The following case, related by Dr. Yeates,* is very interesting, and would be complete if the ultimate report of recovery were of a later date. George E., æt. 20, had been subject to epilepsy for ten years, at intervals of a month; it followed a blow from a pick such as is used by labourers in excavating, which caused considerable depression of the right parietal bone. The depressed portion was removed by a trephine, the opening being two inches long and three quarters of an inch broad; the pulse, which before the operation was small, now became fuller, and the laborious respiration, which had also existed, gradually subsided; the comatose condition remained, and he had seven convulsions during the afternoon and night. Consciousness returned slowly; on the fourth day from the operation sensibility was perfectly restored and the wound was healing. The patient continued to improve and had no return of fits up to the time of the report,

---

* *American Journal of Medical Sciences*, Jan. 1860.

which, however, was only four weeks later. The wound in the scalp was then nearly closed. On one of the buttons of bone removed by the trephine, there was a toothlike process proceeding from and perfectly attached to the piece, about one inch in length, which the writer regards—and probably with justice—as the cause of the convulsions.

Mr. Travers relates a successful case of trephining in epilepsy, which was performed on a lad in whom a depression of the cranium existed. An amusing one is given by Tissot, in which, however, the operation was not performed *selon les règles de l'art*. A young Frenchman who, being affected with epilepsy, went to Italy to be cured, was attacked by robbers, and received, among others, a wound on his forehead which carried off a large portion of the bone; the wound was long open, but got well, and at the same time the patient was cured of his disease.

I have imperceptibly encroached upon the domain of the radical treatment of epilepsy, the treatment to be adopted during the free intervals, and which must be directed to two main points. One indication is necessarily to remove any and every exciting cause capable of producing the paroxysm; the other is to allay, and if possible to banish, that excitability of the nervous centres which enables the irritating cause to produce the fit.

The belief in the powers of medicine over this disease has fluctuated much; and especially do we find the scepticism as to the possibility of controlling the disease to prevail among physicians who have made mental diseases an exclusive study, and have had peculiar opportunities of seeing epilepsy in its most developed form. Esquirol expresses this scepticism formally when he says of epilepsy, "Je n'ai pu obtenir de guérison." I would observe that a lunatic asylum is generally made the ultimate resort of epileptic patients in whom the usual remedies have been exhausted, and in whom incipient mental fatuity has already indicated organic intracranial lesion. For my own part, I should as little wish to send an epileptic into a lunatic asylum for the purpose of cure as I should consider a hospital for consumption a suitable place for a person labouring under incipient pulmonary phthisis. However high in either case the medical talent which presided over the respective institutions, I cannot but think that the congregation of similar cases of disease, and the necessarily depressing effect of being able to compare one's own symptoms with those of surrounding patients, must, especially in such diseases as those adverted to, exercise a baneful influence.

I may premise my further remarks on the treatment of the disease by the observation that I by no means agree with Esquirol. But while I maintain that medicine (and in that term I comprise everything that comes within the domain of the healing art) is capable of effecting much, I am willing to admit that the solitary occurrence of an epileptic seizure indicates a peculiar temperament, a peculiar nervous diathesis, which would always render the individual prone to a return of the complaint. This, however, is scarcely distinctive of epilepsy, but may be regarded as an attribute of almost any derangement in the system. A person who has once had gout rarely escapes a second attack; rheumatism is characterized by the constant recurrence of its symptoms when it has once laid hold upon a victim; sore throats, neuralgiæ, dyspepsiæ, catarrhs of all kinds, are each more liable to recur in the individual in whom they have once made their appearance, than other affections. This is so much the case, that those diseases which ordinarily occur but once in a lifetime are regarded as presenting a type by themselves. Epilepsy only obeys the general law. A person, therefore, in whom this spasmodic affection, or a tendency to it, has once been manifested, deserves the special watchfulness of his medical friend; and the treatment of other affections by which he may be attacked should be regulated by a consideration of the possibility of a reproduction of the epilepsy. But it does not follow that this proclivity renders a return of the complaint necessary, or that there is any other connexion between an attack occurring after the interval of many years and its predecessor, than that implied by the nervous diathesis spoken of. It would not, I think, be philosophical to regard such recurrence as a proof of the actual persistence of the same morbific agent, the same predisposing or exciting causes which operated on the occurrence of the first attack. One of my cases that I have set down as cures had an epileptic fit seventeen years previously to coming under my care. Would it be just to say that the result of treatment in the first attack was fallacious, and did not deserve the name of a cure? Or was I wrong in regarding the case of R. P. as a cure, in whom fits had occurred in infancy, who became epileptic at sixteen, had had weekly seizures before he came under my care, then lost them for above a year, when he again consulted me on account of their return, but lost them again under the employment of the same remedies? In many instances there is no doubt that the result of treatment is merely a temporary arrest or postponement of the affection, which has been erroneously regarded as a cure. Though this result may not be all that is desired, it still is better than if, by allowing the disease to take its course unchecked, we hasten the *facilis descensus Averni*.

We all know how difficult of treatment diseases become in which a moral principle is involved. Fear, love, hatred, anxiety, hope deferred, disappointment, grief, have at least as much to do with illness as physical causes; and certainly much more than is opined by those who seek for all their remedies in pharmacopœias and dispensatories alone. The moral element may not be overlooked in the treatment of epilepsy; and while I utterly disapprove and condemn the employment of falsehood and subterfuge in the treatment of disease, I think it impossible that success can be hoped for in the treatment of the disease by any medical man who does not himself believe that medical treatment may effect much. Hope and confidence are essential elements both in the physician and in the patient. How much such a frame of mind affects patients is shown even in Esquirol's account of the experiments made with new remedies at the Salpétrière; for he says that a new mode of treatment invariably suspended the attacks for a fortnight in some, in others for a month, in others again for two, and occasionally even for three months. After this period, the attacks recurred with their former frequency. I so far agree with the French author, that the value of each individual remedy does not appear to have been established by the experiments, but it is impossible not to admit the effect of hope upon the unfortunate sufferers, and the control which this moral influence exercised over the paroxysms. This lesson is one that Esquirol's remarks teach, and that the

s

medical man may easily acquire for himself; but his experiments do not prove the uselessness of all medication, because the promiscuous mode of administration adopted would in itself be a bar to its success. Confidence begets confidence; and if the physician feels it in himself, he will probably beget it in his patient. Unless he does so, he is not likely to obtain successful results. This moral element should never be lost sight of in the treatment of disease, but least of all in a disease of the nervous system like epilepsy. There are no manœuvres or tricks to be learned. A thorough knowledge of his art, and a proper reliance on the powers at his command, on the part of the Asklepiad, are the bases upon which the intercourse between patient and physician should rest; the tact of the individual must do the remainder, but for that it would be impossible to lay down specific rules. Suffice it to say that, in a patient who has passed the age of infancy, everything that can be done, consistently with justice and propriety, to raise the *morale*, to strengthen the will, to rouse the moral energies, comes within the sphere of the physician. His duty is to treat the patient in all his relations; to have a regard to his moral, his intellectual, his physical nature. Each of these elements is more or less involved in the disease: the physician's tact will consist in determining the relative influence which each exerts; his duty will be to administer his curative agents in such wise as to neglect no one indication. The overwrought intellectual powers will be the main source of the evil in one; the disappointed longings of a yearning heart may deserve special attention in a second; while in a third, a tænia or a blow on the head may be the points to which our attention is chiefly directed. In a sense

PRELIMINARY INQUIRIES. 259

different from the one in which it was originally used, the physician who seeks to treat epilepsy may use as a motto the Terentian

<blockquote>Nil humani a me alienum puto.</blockquote>

Before selecting the mode of treatment to be pursued in an individual case, a most minute and searching inquiry into the patient's antecedents is necessary. Without a full knowledge of the patient's history, the proceedings will necessarily be empirical. We may fail in discovering any basis for rational proceedings, but we certainly cannot avoid sheer empiricism unless we adopt the initiative inquiry suggested. We have seen that a great variety of circumstances may conduce to excite epilepsy: they cannot be surmised. We know of no specific for the disease, like quina for ague, or arsenic for scaly forms of skin disease; but we do know that spasmodic action is the result of a variety of influences to which the nervous system may be subjected, and that, by avoiding or counteracting those influences, it may be held in subjection.

The hereditary predisposition to cerebral disease, and especially to epilepsy, the presence of scrofula and its compeer phthisis in members of the family, must be inquired into; the evidence of the presence or absence of these affections in the individual in early life or at the time of the consultation, is no less important; and all the various influences or exciting causes to which we have drawn attention in an earlier part of this work must be passed in review and eliminated until we have either determined that a certain "cause" exerts such an influence as to merit being singled out for attack, or until we have exhausted our stock of interrogatories. Not till then are we justified in prescribing one of the

s 2

innumerable anti-epileptic remedies for which we possess no special indications.

We may paraphrase Lord Palmerston's definition of dirt, and say of spasm that it is force wrongly applied. We have to see that the force is regulated; that it is generated of proper quantity and quality, and that it is not wasted in efforts for which the individual is not suited.

After the general preliminary examination into the history of the patient, it will be necessary *seriatim* to examine the state of individual organs. The head, its coverings, the evidences of local injury, the temperature of its surface, may first be inquired into. The state of the digestive organs will command special attention, as well as the genito-urinary organs. The vascular and respiratory systems will necessarily be inquired into also. The state of the cutaneous coverings, the presence or absence of skin affections; the distribution of temperature; the complexion, as indicating a plethoric or an anæmic condition, will successively be passed in review, and their condition noted. Not till all this is done should the physician attempt to arrive at a conclusion as to the nature of the case before him.

If I were to formularize the prevailing mode of treatment which I myself adopt, I should say it consisted in local derivation,* or counter-irritation directed against cerebral congestion, and, in general roborants or tonics; the selection of the special mode in which the latter indication has to be carried out necessarily

---

* Tissot goes so far as to say: Tout ce qui peut augmenter la quantité du sang ou la déterminer à se porter plus abondamment à la tête, doit occasioner l'épilepsie.

depending upon the results of the inquiries into the condition of the individual organs.

The great prevalence of headache in epilepsy, either as an habitual affection, as a precursor of the attack, or as a sequel, accompanied by other symptoms indicating a congestion to the head, suggests the propriety of attacking the head, according to the intensity of this symptom, by more or less severe counter-irritants. These vary in strength, in their rapidity of action, and in the effect which they produce upon the system. We employ chiefly blisters, the ointment of tartrate of antimony, setons, and the actual cautery. The first agent acts rapidly and superficially, but in many cases suffices to remove temporary symptoms of headache and vertigo; yet it would answer our purpose where the vertigo is a premonitory symptom, because some hours are always necessary to produce an effect; then dry or wet cupping must be had resort to. Tartrate of antimony ointment takes a still longer period to produce the necessary pustulation, the number of inunctions required varying in different individuals according to the character of their skin or their surface circulation. It is a most valuable agent in many cases, especially where we have reason to attribute the disease to anything of a metastatic character, such as suppression of a cutaneous eruption. It never produces toxic symptoms when applied epidermically, the effect being limited to the point at which it is applied. Setons, again, produce a more permanent effect, and by causing a persistent purulent discharge, keep up a counter-irritation, which is of the most undoubted use in long-standing epilepsy, accompanied by well-marked head symptoms. The effect has repeatedly been so marked as to render the reapplication of the seton necessary where the threads had been accidentally or intentionally removed. While writing these lines I have two severe cases under my care which palpably illustrate their utility. Lastly, the actual cautery, though an agent more employed abroad than with us, deserves great commendation; whether acting by reflex or by the actual drain set up, it sometimes appears to produce curative effects which could scarcely be expected from milder remedies. Now that the use of chloroform prevents the pain of the application being felt, we may hope that the *ferrum candens*, which has almost fallen into disuse in England, may come into more general favour. It must be applied in the course of the spine in such a way as to produce a sufficient eschar, the separation of which insures a powerful counter-irritant effect, which is kept up by the discharge accompanying it. But even if we are unable to cure, we may by counter-irritation often materially alleviate the symptoms. I have repeatedly earned the thanks of my patients whom I was unable to cure of their epilepsy, by relieving them of the intense cephalalgia of which they complained; but I am also satisfied that in some of my cases the cure* was materially aided by the application of setons and other counter-irritants.

The following cases may be quoted in illustration of the previous remarks:—

C. D., aged 11, the son of a gardener, five or six years before the consultation (April 30, 1848), had an "abscess in the head," and otorrhœa; since which time he has been subject to occasional fits of uncon-

---

* In speaking of a cure of epilepsy, I always mean an "apparent" cure, fully appreciating the difficulty of determining whether it is radical or not.

sciousness, which seize him generally without premonitory symptoms; at times he is previously troubled with giddiness, and if he then lies down the attack will pass off. The unconsciousness lasts from five to ten minutes; and on recovery there is a sense of weakness, which lasts for half an hour. He neither screams, bites his tongue, nor is convulsed. He has frequent headaches and of late occasional sickness, but is otherwise well. The fits recently about once a fortnight, formerly not so frequent. Head occasionally hot. No evidence of heart disease, or gastric disturbance. With the exception of a little castor-oil and an alkaline diuretic, the treatment consisted in a seton in the neck, which he wore for nearly six weeks. The attacks did not return; and in 1853, when his brother came under my care for a similar affection, there had been no relapse.

E. W., an unmarried woman, æt. 50, had been subject to epilepsy for seven or eight years, but the attacks had been worse for the last four or five, occurring indifferently at all times of the day and night. She had always been regular till the catamenia ceased about seven or eight years ago. Since then the abdomen had enlarged, but she had not suffered from leucorrhœa or piles. She bites her tongue much in the fits. The bowels open, pulse 88. There is much vertigo, and constant cephalalgia.

Oct. 17, 1859. A seton was applied to the nape, and a quarter of a grain of extract of belladonna given three times daily. To promote the discharge the unguentum sabinæ was subsequently rubbed over the seton. On the 25th Oct. the head was reported better; on the 8th Nov. she stated that she felt well. The pills were exchanged on the 25th for a mixture of dec. cinchonæ and iodide of potassium, with which she persevered to Jan. 6, 1860, when no fresh attacks having occurred, and the vertigo and headache not having returned, she was discharged "cured."

In the case of G. C. K., a married tailor, fits had occurred for two years at short intervals, an aura mounting upwards from the stomach; he had no headache, but owing to the disappearance of an acne on the back, there seemed to be a special indication for the employment of counter-irritation by ung. antimon. potassio-tartratis. The patient experienced immediate relief repeatedly on the appearance of the pustular eruption; and for this reason a seton was at last applied, and ordered to be worn in the neck, from which he seemed to derive permanent benefit.

Among the instances spoken of previously as acute epilepsy was one to which I said it would be necessary to refer in discussing the treatment. Whether it was a fortunate coincidence or not, it deserves mention that the actual cautery having been applied at the wish of Dr. Gueneau de Mussy, the attending physician, a cure resulted. It may be remembered that decided epileptic fits supervened suddenly in a gentleman liable to rheumatic attacks. In the course of three days, before the consultation, he had six attacks by night and by day; no local affection could be traced except a slight tenderness of the spine in the upper dorsal and lumbar regions. On the assumption that we had to do with a rheumatic affection of the spinal meninges, we ordered iodide of potassium and the actual cautery to the nape of the neck, where it was very effectually applied by Mr. A. G. Lawrence, now of Chepstow. In the course of the succeeding forty-eight hours the patient had five more attacks, but slighter than the previous ones; and after this period no further return took place.

When we wish only temporarily to produce a counter-irritant effect, the application of an occasional blister may suffice; it is regarded with less fear by the patient, and if kept open by savine ointment, answers very nearly the same purpose as the issue. But wherever it is important to secure a permanent vicarious discharge, the latter has the advantage of producing a profounder effect upon the economy.

When the symptoms of congestion to the head have been of a temporary character, dry cupping affords relief, and but rarely is it necessary to have recourse to the scarificator. As far as my personal experience goes, I should join with those who deprecate the abstraction of blood in epileptic subjects for any other purpose than that of derivation, as for instance in the case of suppressed menstruation, where it is at times invaluable. Still, it is probable that, among the sanguineous patients that a country practitioner is likely to meet with, venæsection may at times be practised with benefit. The evidence in favour of that proceeding in some, especially British, writers, is such as to compel the admission that it is at times highly beneficial. Thus Dr. Cooke, among the moderns, advocates venæsection, leeches, and purging as the treatment to be pursued in plethoric cases; and the method appears to have been successful in his hands. From the very opposite statements by different authors on this point, we cannot but conclude that the disease presents a different aspect at different times and localities. This is not merely a charitable way of accounting for doctors' disagreements, but consonant with the experience afforded in daily life of the variations in the forms of disease.

The effect of compressing the carotids may fairly be assumed, so far as regards the brain, to be analogous to depletion; this proceeding can, however, scarcely do more than produce a temporary effect. To render the diminished vascular tension, thus obtained, permanent, it would be necessary to ligature the carotids; and this has actually been done. Compression is not, however, a remedy which is often successful in averting or avoiding the fit. We learn from Dr. Delasiauve's recent Treatise of Epilepsy,* as well as from Dr. Burrows,† that ligature of the carotids has been employed in several instances with advantage. Both refer to a case in which Dr. Preston, in Calcutta, tied the carotid artery, the operation being followed by temporary success. Delasiauve mentions that an epileptic subject having cut his thyroid artery, with a view to suicide, M. Boileau tied the carotid, in consequence of which both the hæmorrhage and the epilepsy were stayed; he also states that Velpeau was accidentally called to tie the temporal and facial arteries in an individual subject to daily attacks, who was cured by the operation. Dr. Delasiauve observes that, though these results are seductive, they do not suffice to authorize the performance of so serious an operation. Still, they appear to justify our not regarding Kussmaul and Tenner's

---

* Traité de l'Epilepsie, p. 426. Par le Docteur Delasiauve. Paris, 1854.
In a paper by Dr. Wood, on Ligature of the Common Carotid Artery (*New York Journal of Medicine*, new series, vol. iii. pp. 9-64; and Mr. Chatto's Surgical Report, *British and Foreign Medico-Chirurgical Review*, Oct., 1857, p. 543) this operation is stated to have been performed twice in New York for epilepsy; both cases were benefited but not cured.

† Disorders of the Cerebral Circulation, 1846, pp. 72-79.

conclusions with reference to the causation of epilepsy as universally applicable, and rather to encourage the view that various and opposite states of the circulation may in different persons induce the disease.

As a general rule, the treatment of epilepsy may be commenced with a single brisk purge, in order to make sure of the primæ viæ being properly emptied, and allow full play room for the further treatment. The presence of worms, of hard scybala, is often not suspected by the patient; and their removal is a *sine quâ non* of success. Still it will be better not to follow a mere empirical rule, but to seek for the rational indications of a given agent.

The employment of purgative medicines may be indicated in epilepsy for various purposes. The form in which they are exhibited will necessarily vary accordingly. The chief indications may be summed up under the following four heads:—They are given to secure the normal evacuation of waste matter; to derive from the head; to expel foreign matters or worms lodging in the intestines; or to promote certain physiological secretions. It is consonant with our general view of epilepsy that the purgatives employed during its course should not be of a character to impair the plasticity or diminish the respiratory functions of the blood. Drastic purgatives, purgatives that produce a very lowering and depressing effect, should therefore, as a general rule, be avoided, and those selected that are of a warm aromatic character. To secure the first indication, of simply regulating the bowels, as it is called, it will often suffice to make a temporary alteration in the diet,—even a change in the dinner-hour, and a change of beverage, may secure the object; in many instances the omission of tea or coffee, and the substitution of milk or cocoa for the morning and evening meal will be desirable on this ground; the habitual exhibition of the mildest purgative medicine keeps up an irritant action in the alvine viscera, which we too often see fraught with the most baneful consequences.

The temporary relief afforded by the evacuation covers the pernicious effects of the bad habit, which are demonstrated in the generation of a host of symptoms indicating an enfeebled nervous system. The costiveness of many of our overworked and anæmic patients will be better met by a large dose of quina, or by nux vomica or its alkaloid, than by a *haustus purgans*. In some cases, possibly, even an opiate or a sedative will more readily induce a regular action of the bowels, by overcoming a spastic condition of the intestinal muscular fibre. I would especially enter my protest against mercurials, which can scarcely ever be required on the vague ground of "improving the secretions;" their use ought to be much restricted in the treatment of disease generally, but in epileptic conditions they ought not to be administered without the most definite indications,* and should on no account be given so as

---

* Marshall Hall (On the Neck as a Medical Region. Essay sixth. 1849) says in reference to the employment of mercurials, "Especially the system should, I think, be kept under long-sustained mercurial influence. From this influence, if conjoined with exercise in the open air, no harm whatever results, and from some extrordinary cases, I am perfectly satisfied that it does infinite good, mitigating the number and force of the attacks, and in due time—that is, a very long time—subduing their effects." The writer gives no specific indications for the employment of mercury, and relates no cases by which we can judge of the results of his treatment; we cannot, therefore, adopt his αὐτός ἔφη as a sufficient ground for abandoning the views expressed above.

to "affect the system." Rhubarb, the compound colocynth pill, aloes, castor-oil, taraxacum, sulphur in combination with magnesia or rhubarb, are among the laxatives that are most suited to epileptic subjects. Nor may I omit the mention of a mineral water, which is coming into general use in England, and possesses admirable qualities as a laxative. I allude to the Pullna bitter water, which is imported from Bohemia, and owes its properties mainly to the sulphate of soda, which it contains in the proportion of about one hundred grains to the pint. It contains not much less of sulphate of magnesia, with smaller quantities of sulphates of potash and lime, carbonates of lime and magnesia, chloride of magnesium, and phosphate of lime, with free carbonic acid. The whole amount of saline constituents is about two hundred grains to the pint. Half an ordinary tumbler of water, taken in the morning, generally suffices to produce a full, pultaceous evacuation. In delicate subjects it is well to give it with equal portions of warm milk. The waters of Friedrichshall, which are also imported into England, are similar in their effect to those of Pullna, and both have the advantage of not being followed by the costiveness that so often embarrasses the physician in the employment of ordinary drugs. I have often found habitual costiveness relieved permanently by them, the dose being gradually reduced to a minimum.

Among purgatives we must advert specially to turpentine; a remedy which may often be had recourse to, both on account of its primary action upon the intestinal canal and its secondary stimulant effects in disorders of the sexual system. In the variety of epilepsy associated with an hysterical constitution, it is often valuable. Many physicians have employed it, but none appear to have been so successful in their choice of this agent as Dr. Prichard, who details numerous very satisfactory cures of epilepsy by means of turpentine. Dr. Watson also speaks of turpentine in terms of high praise. I have not enjoyed the same measure of success in my use of this drug; a circumstance which I can only explain by assuming that the constitutions of the patients to whom I administered it differed from those of Dr. Prichard's; in the same way as the venæsections which were beneficial in his hands and in the hands of several of the older physicians of eminence, are repudiated by practitioners of the present day.

Nor can I think that we are justified in arrogating to ourselves so much superior tact and knowledge in the treatment of disease generally as to decry all that has been done by our predecessors, though it clashes with many of our views. When we read the careful histories that they have handed down to us, it is often impossible not to be struck with the masterly manner in which they handled their tools. The remark applies forcibly to a disease like epilepsy; the advantages supplied by "physical diagnosis" have afforded us no means of elucidating the affection which was not possessed by the writers alluded to.

I cannot hesitate to admit that Prichard, Cooke, and others have found venæsection an important auxiliary—in some cases the sheet anchor—in the treatment of epilepsy, although I have hitherto not met with cases in which I should be disposed to employ it. The prevailing character of the pulse during the free interval in my cases was feeble, indicating anæmia rather than plethora, and demanding an infusion of new, healthy blood, rather than a diminution of the small current taking its sluggish or petulant course through the vessels. The employment of leeches in small numbers is indicated where we desire to draw away the blood from a part, rather than diminish the general tension of the vascular system; thus in persistent congestive headache their application to the nares or temples; as an adjuvant to the restoration of the catamenia, their employment at the perinæum or hypogastrium, is often valuable.

The prevailing opinion among writers of the present day is that anything like heroic antiphlogistic treatment in epilepsy and epileptiform disease generally ought to be eschewed,—an opinion that I cordially adopt; the drugs pertaining to that category ought to be so used as to restore order where disorder prevailed; to rectify the vitiated secretions where they can be shown to be deranged; to remove local congestions or other accumulations where such means suffice for the purpose.

While, then, I do not deny that epilepsy may be the result of too high a pressure, the evidence proves it in the vast majority of instances which we meet with in the present day to be due to a want of steam—of more pressure. Accordingly, the remedies most in repute in the treatment of epilepsy are those which are commonly classed together as tonics; and among these we find especially the mineral tonics to deserve and to hold a high rank. Drawing the circle still narrower, I should be disposed to place the preparations of iron and zinc first, as those which have done me most service.

The various salts of iron may be given according to the different constitutions of our patients; but generally the vegetable salts, the ammonio-citrate, the potassio-tartrate of iron, the ferrum pomatum (a malate) of the Prussian pharmacopœia, with which we may class the lactate, are preferable, on account of the facility with which they are digested. Where there is want of appetite, the citrate of iron and quina is a very appropriate form of administering iron. The irritant properties of the sulphate of iron render it generally ill-suited. A very elegant form of administering iron, and one that is particularly well adapted for young children, is under the guise of Allarton's steel biscuits, which are most palatable, so as to be eagerly taken, even by the infant.

Among numerous cases that I might bring forward to illustrate the above, I select the following :—

Miss N. O., an unmarried lady, æt. 27, of spare habit, florid complexion, and nervous disposition, consulted me in March, 1858. She had always enjoyed good health till a year previously, when she suddenly one evening after supper fell down in a fit. Since then the attacks have recurred with increasing frequency, generally once a month, lately oftener. The attacks have no relation to the catamenia, which are regular. The fits come on without premonition, and are marked by entire unconsciousness, cramped hands, a bitten tongue, and extreme pallor. There is no lividity, but the neck swells and the surface veins become prominent. Recently she has had almost continuous headache, which is aggravated by the fits. There is no hereditary or other dyscrasia traceable. She is habitually subject to chilly feet, the bowels are sluggish, the urine normal, containing no albumen or sugar and exhibiting no deposit. There is occasional *petit mal.* She was ordered nutritious, easily digestible food, with an avoidance of salt meat, pastry and wine, feather beds and feather pillows. Regular hours and baysalt sponging were enjoined, with the following medicines.

℞ Strychniæ gr. j.
　Extr. gentian. gr. lxxiv.
　Extr. aloes gr. vj.
　Fiat massa in pilulas xvj. dividenda.
Sumat pilulam unam ter die.
℞ Linim. camphoræ.
　Linim. terebinth. āā ʒj.
S. mane et nocte dorso infricandum.

With slight variations this was continued to April 1, when it was necessary, from symptoms of trismus, to omit the strychnia; otherwise she was doing well.

℞ Ferri et quinæ citr. ʒj.
　Syrupi ʒj.
　Aquæ cinnam. ʒiij.
Sumat cochl. j. amplum ter die.
℞ Pilul. hydrarg. gr. j.
　Extr. rhei gr. iij.
　Confect. arom. gr. j.
Alterna nocte sumend.

April 14. Reported herself to be doing well, to have no headache, and no return of fits.

℞ Tr. ferri mur.
　Ethr. chlor. āā ʒiss.
　Syrup. zinzib. ʒvj.
　Aquæ ad ʒiv.
Sumat ʒss. ter die.
℞ Sodæ potassio-tart. ʒss.
　Sodæ bicarb. ʒj.
　Extr. glycyrh. gr. xx.
　Ether. chlor. ♏xx.
　Inf. rhei ʒiij.
S. dimidium bis hebdomade sumendum.

April 20. She was so well that medicines were discontinued, but the cold sponging ordered to be persevered with. On May 3 she had an attack; soon after she was ordered the valerianate of zinc in pills, with Pullna water and milk every morning to obviate the costiveness. The potassio-tartrate of antimony was also ordered to be rubbed into the neck for a time on account of a return of headache.

May 27. Is much stronger, and has no pain.

℞ Ferri valerian. gr. xlviij.
 Extr. absinthii ʒss.
 Ol. menthæ ♏vj.
 Fiat massa in pilulas xxiv. dividenda. Sumat
  binas ter die.

June 12. Three days ago she had brief temporary unconsciousness, not amounting to a fit. There was slight tendency to bleeding of the gums, and some purpura.

℞ Ammon. carb. ℈viij.
 Ferri citr. ℈iv.
 Syrupi ʒj.
 Aquæ sambuci ʒviij.
 Sumat unciam ter die cum succi limonum
  ʒij. sub effervescentia.
 Haustus laxans cum phosphate sodæ.

From this time no further bad symptoms occurred, and the lady ceased her attendance. I heard, however, in January, 1859, from the mother, that she remained perfectly well, and I have reason to believe that, had the fits returned since then, I should have been informed.

J. L., the son of a labourer, æt. 14, consulted me some years ago, on the 30th Dec., on account of epilepsy. He had had no fits in infancy, and with the exception of measles, smallpox, and hooping-cough, had always enjoyed good health. Two years previously he was first attacked with fits when engaged in paper-staining. He was seized suddenly, without premonitory symptoms. He does not scream, but moans, and bites his tongue and lips. After the fit he is sleepy and giddy. He has no headache, but at times experiences a drumming in his head. Altogether he has had seven attacks. Recently they have become more frequent; he had one on Dec. 23rd, and another on Christmas-day. They occur at different times of the day. The tongue is clean, the appetite good; the pulse sixty-eight, of good strength. No worms have been seen, but they are suspected.

   ℞ Hydrarg. chlorid. gr. iv.
    Pulv. jalapæ gr. viij.
    Pulvis statim sumendus.
   ℞ Tr. ferri muriat.
    Ether. chlor. āā ʒiv.
    Aquæ pimentæ ad ʒvj.
    Sumat ʒij. ter die ex aquâ.

Jan. 2. Feels and looks better; urine sp. gr. 1022, without sediment, albumen, or sugar. He continued to progress favourably, and having had no return of the fit up to Feb. 3, he was discharged, apparently in perfect health.

Of zinc I would speak very favourably, though by no means with the confidence of M. Herpin. I have satisfied myself again and again that it exercises a distinct influence over the epileptic paroxysm, often inducing an entire cessation, though frequently only causing a postponement of the attacks. I much prefer the soluble sulphate to the insoluble oxide. The former affords us an instance proving the extreme power of the system in adapting itself to hostile impressions, if we may say so, provided the attack be gradually made. To a person in health, five grains of the sulphate taken at once are liable to prove emetic; but by cautiously increasing the dose, epileptic patients can be brought to take more than seven times that quantity repeatedly in the day, with beneficial results. One of my patients, in whom the dose was gradually augmented to thirty-six grains three times a day, appeared to be completely cured of his paroxysms. I never doubted the fact of his taking the medicine, as he assured me he did, conscientiously; but I satisfied myself by requiring him to swallow a dose in my presence, that thirty grains had no unpleasant effect upon him whatever. The zinc may be given in pills with extract of gentian, or in infusion of valerian, or other combinations indicated by the particular case. The oxide of zinc is a very insoluble substance, and does not seem to possess as energetic properties as its relative the sulphate. It has not in my hands proved equally satisfactory. The valerianate of zinc and the valerianate of iron present combinations of the bases spoken of with valerianic acid, which may be given with advantage. I commonly combine small quantities of rhubarb with the zinc, to counteract the astringent properties which it exerts on the intestinal canal. It has not been ascertained what becomes of the zinc. That it is actually introduced into the system and deposited in different parts can scarcely be doubted by any one who has watched its effects, though there is no visible evidence of such a result as there is in silver. It does not appear to pass out by the kidneys, certainly not in any appreciable quantity; I am not aware of any analysis having been made to determine the amount contained in the fæces. A case is on record* of a gentleman having taken in the course of five months

---

* *Brit. and For. Med. Rev.* July, 1838, p. 221.

3246 grains of the oxide of zinc, in consequence of which he wasted away and was brought to the brink of the grave. Such effects are so extremely rare that they need scarcely be taken into account, for any intolerance of zinc is commonly manifested by vomiting. I shall shortly give a case in which nearly six times the above amount of sulphate of zinc was given without injurious effects. As long as it agrees with the patient—and he commonly improves much in appearance under a suitable dose—he may continue it, and as soon as symptoms of intolerance are manifested, it must either be omitted, or the dose diminished. It does not possess a cumulative character.

The following cases are adduced to demonstrate the utility of the salts of zinc in the treatment of chronic epilepsy :—

R. P., a clerk, æt. 17, of a scrofulous appearance, dark complexion, and rather undersized, had had fits during the period of dentition, and had been subject to headache all his life. An uncle had fits. The first epileptic seizure occurred when he was sixteen; a second took place four months later; since then they have returned weekly. He screams, is unconscious for about an hour, has bitten his tongue twice. The headache is urgent before the fits, and there is a momentary warning of their actual approach. There is no albumen or sugar in the urine. From April 21, 1854, to May 9, he took sulphate of magnesia drafts and cotyledon umbilicus. From May 9 to July 25, he took sulphate of zinc, gradually increasing the dose from two to ten grains three times a day, in infusion of calumba. Between April 21 and May 9, he had two fits, though less severe than the previous ones. After this there was no recurrence of fits; and on the 25th July he was discharged cured. A year later the same person again came under my care, on account of a return of the paroxysm, April 10, 1855.

℞ Valerian. zinc. gr. j.
Extr. gentian. gr. iv.
M. fiat pilula ter die sumenda.
Sumat binas pro dosi tertio ab hinc die.

April 17. One slight fit six days ago. Sumat pilulas iij. ter die, et postea quatuor.

April 27. No fit.

℞ Zinci sulph. ℨij.
Inf. valerian. ℨxij.
Sumat ʒss. ter die.

This mixture was continued regularly up to June 8, when, no fit having recurred for two months, he was discharged cured, and he has not since returned.

B. C., æt. 20, an unmarried gardener, of an anæmic appearance and slight build, never had any serious illness in his life, and never had fits till the present seizures came on two years ago. The first came on suddenly without a warning, as he was coming down stairs. He is not subject to headaches, though occasionally there is slight headache after the attacks; lately they have been preceded by a sense of strangeness. They only last a few minutes; he does not generally scream, but bites his tongue in the fits. He sleeps after they are over. The appetite is good, tongue white, pulse 60. No definite cause was assigned, but he admitted himself to have been guilty of masturbation before the attacks came on. There was no albumen or sugar in the urine. He was ordered, Feb. 6, 1855, to employ cold sponging, to take cotyledon umbilicus (ʒss ter die), and pil. rhei 10 gr. ter die. Up to April 10 he had three fits. Sulphate of zinc pills

(gr. ij. pro dosi), were then given three times a day.
April 20.

 ℞ Sulph. zinc. gr. ij.
  Aloes gr. j.
  Sumat pilulas binas ter die.

April 27. One fit rather sooner than usual.

 ℞ Sulph. zinc. gr. ij.
  Extr. gent. gr. iij.
  Sumat pilulas tales tres ter die.
  Habeat haustum magnesiæ sulphat. cum
   magn. carb. alterno die.

May 11. No fit. Sumat pilulas quatuor pro dosi et augeatur dosis pilul. j. quarto quoque die. Rep. mist. May 25. No fit; a longer interval than usual. Rep. pilul. quatuor pro dosi. June 12. One slight fit a week after last visit.

 ℞ Sulph. zinc. gr. x.
  Inf. valerian. ʒss.
  Ter die sumend.
 ℞ Pil. rhei. co. gr. x.
  Bis hebd.

June 26. No fit; no sickness from medicine. Adde sulph. zinc. gr. iv. sing. dosibus. July 13. No fit; no sickness or nausea after medicine. It is a fortnight over the usual interval. Bowels open; good appetite. Rep. mist. adde sulph. zinc. gr. ij. (gr. xvi. pro dosi). July 27. Had a fit a week ago of the same strength as the previous one, the longest free interval since the supervention of the epilepsy. There was no particular cause. He has felt rather sick after some doses of the last medicine. It was, however, persevered with in the last dose; and the next free interval was six weeks. Sept. 11. Rep. mist. adde sulph. zinc. gr. ii. (Dosis gr. xviii.) Habeat pil col. co. gr. x. bis hebdom. Sept. 28. Feels well, has had no fit; twice felt sick, but did not vomit; the dose was increased to twenty grains, with which he persevered to the 30th Oct., when it was increased to twenty-two grains; on the 13th Nov. it was increased to twenty-four grains. Nov. 30. Has taken the medicine regularly, experiencing no sickness, and only slight nausea in the morning. Pergat. Dec. 18. No fit now for seven weeks; he feels no inconvenience from the medicine, and takes it regularly. The urine contains no albumen or sugar, and no precipitate is formed by ammonia. The dose increased to twenty-six grains. Jan. 1, 1856. No fit for nine weeks; is looking very well; dose increased to gr. xxviii. Jan. 28. No fit for thirteen weeks; has taken the mixture regularly, not having missed once. Once or twice, early in the morning, he has been sick. Appetite excellent. He feels generally stronger than formerly. Dose increased to thirty grains, three times a day. Rep. pil. coloc. co. Feb. 15. No fit now for fifteen weeks; takes his medicine regularly without any inconvenience. Dose increased to thirty-two grains. Feb. 29. No fit. Mr. Copney, then the dispenser of St. Mary's Hospital, an excellent analytical chemist, carefully examined the urine, and could detect no zinc in it. Dose increased to thirty-four grains. March 14. No return of fits for about five months. He feels strong and well; experiences no sickness after medicine, which he takes regularly. The bowels are open, the pulse 68, full and large. Dose increased to thirty-six grains. After this time he ceased attending.

It appears that this patient took from the 10th April to the following 14th March, 18,976 grains of the sulphate of zinc, amounting to 3lbs. 3 oz. 6 dr.

MINERAL TONICS. 281

16 grains. During the whole time there was no intermission, and none but the most beneficial results were observed. The sulphate of zinc cannot, therefore, be regarded as a poisonous agent.

These two cases speak for themselves. I am, however, far from asserting the universality of the salts of zinc in this affection. Why they should be so useful in some cases, and fail in others, must depend upon a peculiarity in the morbid changes inducing the fits which yet remains to be discovered. The following is an instance in which the steady use of the zinc utterly failed in even arresting the fits.

D. C., a single woman, æt. 17, always delicate, had her first fit at seven years of age, after swallowing the core of an apple; the second occurred when she was fourteen; subsequently every fortnight, and for three or four months before she came under treatment, weekly. Catamenia irregular and scanty. The fits last from ten to fifteen minutes; there is complete unconsciousness, with flushing of the head and neck, limited to a horizontal line across the neck. The head and body are twisted to the right side, the right side of the tongue is frequently bitten, and there is a choking in the fit, involuntary micturition, and the pupils are enlarged at the time. The fits occur indifferently by day and night; she sleeps after the fits; there is no aura. The bowels are costive, the appetite good, the pulse calm, the tongue slightly furred. There is great restlessness during the sleep. No albumen or sugar in the urine. From July 1 to July 15, she took amorphous phosphorus in ten-grain doses, three times daily, with compound aloes pills. From July 15 to July 29, the oxide of zinc gradually increased from two and a-half grains to five grains, three times daily, in infusion of valerian. On the 29th July sulphate of zinc was substituted for the oxide, and was given in increasing doses up to twelve grains three times a day, in infusion of valerian. The compound aloes pill and cold sponging were continued throughout. No improvement resulted.

I could add numerous cases illustrative of the benefit obtainable from a judicious administration of salts of zinc. I shall confine myself to briefly noticing one apparently hopeless case, in which a whole month's free interval was obtained in a patient who had rarely been free for more than a day or two.

P. Q., a single lady, æt. 24, spare and anæmic, whom I saw in consultation with Mr. Verral, had been subject to fits from the period of dentition. At the first attack, the right side of the body became paralysed, traces of which still remain in a contraction of the hand and arm, and a somewhat atrophic condition of the whole side. The fits frequently occur every night, and the free intervals never extend beyond forty-eight hours. There is no headache, and the memory has not failed. I pass over the other symptoms, and merely mention that she was ordered bay-salt sponging every morning an hour after breakfast, nutritious diet, with bitter beer, and the following prescriptions:—

℞ Ferri valerian.
Zinc. valerian. āā gr. j.
Extr. gentian. gr. ij.
Fiat pilula ter die sumenda; et tertio quoque die dosis augeatur pilulâ unâ.
℞ Extr. aloes aq. gr. ij.
Extr. hyosc. gr. j.
Extr. absinth. gr. ij.
Fiat pilula alterna nocte sumenda.

I heard by letter that she remained free from fits

from the 6th Jan. to 6th Feb., since which time I have received no further account.

Among tonics we must not forget to mention strychnia, which, in suitable doses, acts as a general roborant, and diminishes that irritability of the nervous system which prevails in persons subject to epileptic seizures. An elegant form of the combination of strychnia and iron is the double citrate, containing one grain of strychnia in a hundred; five grains are the proper dose for an adult. The extract of nux vomica in half-grain doses three times a day, with extract of gentian, acts in a similar way; and it is worthy of remark that we often observe the sluggish state of the bowels associated with epilepsy rectified by the exhibition of the remedies just spoken of in such a manner as to render the administration of direct purgatives unnecessary.

The preparations of silver which have been praised by some as exerting a beneficial influence upon epilepsy, have not answered my expectations, though I would not agree with Georget in entirely erasing them from the list of agents to be employed; the tonic properties of the nitrate and the oxide are scarcely to be denied, though undoubtedly the irritant properties of the former render greater caution necessary than is required where we administer the latter. In a female who died in the Salpêtrière under Georget, and who before admission had taken the nitrate of silver for eight months, the mucous membrane of the stomach was found entirely destroyed at its lower half; the stomach was perforated at four or five points, and nothing but the peritoneal coat remained at several others. Georget* says, in

---

* For a case in which nitrate of silver appears to have cured the epilepsy, but to have brought on ulceration of the stomach,

reference to this case, that he cannot comprehend how any one can be so blind as to expect to cure cerebral

---

see Virchow's "Archiv." Band xvii. Heft 1 & 2, and my Report on Pathology and Medicine, *Med.-Chir. Rev.* April, 1860, p. 544. It occurred in the German Hospital in London, under the care of Dr. Frommann, in a man aged sixty. The patient became epileptic in March, 1856, and was treated with nitrate of silver almost from the commencement; for nine months he took a daily pill containing six grains, so that during that time he swallowed nearly three and a half ounces. Towards the end of July the skin began to be discoloured, but in spite of gastric symptoms the remedy was persevered in. In 1857 hæmatemesis and other symptoms of gastric ulceration supervened, while the severity of the epilepsy had abated, and having in the meantime come to England he was admitted into the German Hospital, where he soon died. The special interest attaching to the autopsy is connected with the extent to which the silver had been deposited in the tissues. The parts in the face which had exhibited the greatest intensity of discoloration, owing to their containing more blood, now presented a tint uniform with the rest. In the brain, the choroid plexuses presented a uniform greyish blue tint. The lungs were tuberculous and pneumonic, the heart hypertrophic. The stomach contained a large quantity of acid brown liquid, streaked with blood, and at the upper part of the posterior wall, halfway between the pylorus and cardiac orifice, was a large ulcer, at the base of which was an orifice blocked up by the adherent pancreas. The mucous membrane of the duodenum and jejunum was dotted over with many small black granules, most closely aggregated along the folds. In the ilium these spots became more and more scanty. Examined by the microscope, the villi in these black spots presented, especially in their globular end, groups of black aggregated particles, varying much in form and size, and without a crystalline character; cyanide of potassium rapidly dissolved these deposits here as well as in the other organs in which they were found. The spleen was small, its veins had an ashen hue, which was due to a finely granular precipitate upon their coats. The liver was small, congested and fatty; the small branches of the vena portæ and of the hepatic veins presented the same precipitate of silver

disease by cauterizing the stomach; it is not probable that any one who prescribes the nitrate of silver does it with this intention.

I can say nothing in favour of copper and its salts. Among the general tonics that are in vogue in the treatment of epilepsy there are few that deserve greater commendation than the mineral acids, and especially the nitromuriatic. Though I have stated my reasons

---

throughout, but the capillaries were free from it. Fine sections of the hepatic tissue showed numerous black dots, each of which occupied the centre of an acinus, corresponding to the point of exit of a central vein, and the colour was produced by a black margin surrounding the calibre of the artery. The dark colour of the branches of the vena portæ was also very characteristic throughout. The largest argentine deposit was in the kidneys, where the bundles of vessels in the Malpighian corpuscles and the intertubular capillaries seemed to be its primary seat. The pyramids all exhibited a dark grey colour, which was deepest and all but black near the papillæ. The tubules in these parts were entirely invested with a dense precipitate, so that on a transverse section each tubule appeared surrounded by a black ring. Parts of the skin taken from the temporal, axillary, and digital regions were examined. Transverse sections showed a pale purplish streak immediately underneath the rete Malpighi, following the undulations of the cutis. At the roots of the hairs it accompanied the external sheath towards the bulb, but nowhere except in the sudoriparous glands was a granular deposit to be found; in them it presented an appearance similar to that seen in the renal tubules. The glandular epithelium uniformly presented fatty degeneration. The deposit was dissolved by concentrated sulphuric acid and cyanide of potassium. An analysis by Dr. Versmann gave the following results: 217 grains of dried liver yielded 0·009 grammes of chloride of silver, 0·0068 grammes of metallic silver, or 0·047 per cent. of metallic silver; 133 grains of dried kidney yielded 0·007 grammes of chloride of silver, or 0·0053 grammes of silver, or 0·061 per cent. of the latter.

for not acceding to Dr. Hunt's* views with regard to the *causa proxima* of epilepsy, I have seen frequent instances in which the exhibition of this tonic was of signal service; derangement of the primæ viæ is a very common accompaniment of ˋepilepsy, and in most of those cases the nitromuriatic acid in water, or with gentian, hop, or similar vegetable bitter, deserves our confidence; when the urine contains a deposit of oxalates, copious pale lithates or phosphates, the indication is particularly urgent. It will often be the intercurrent treatment, to meet particular circumstances; at times it appears to suffice by itself to arrest the paroxysms. I shall only quote a single case in illustration.

M. N., a pale, spare, rather undersized, and precocious boy of nine years old, was brought to me from the country on March 21, 1860, by his father, a surgeon. He had but recently been seized with epilepsy, and had five attacks before I saw him. He had had dentition fits, but none since the scarlet fever at the age of two, till the seizures for which I was consulted. There was no local disease traceable, and no onanism or worms proved. There was general want of tone, loss of appetite, especially for animal food, copious deposit of lithates in the urine, a small pulse of 104, and a clean tongue. He was ordered bay-salt sponging, early rising, a hard bed, and nutritious diet.

℞ Tr. ferri mur. ʒiss.
Tr. belladonnæ ♏xvj.
Syrupi ad ʒij.
M. sumat. ʒj. ter die.
A powder of rhubarb, magnesia, and ginger occasionally.

* *Medical Times and Gazette*, vol. xii. 1856.

The intervals were lengthened out by this treatment, but in the course of a month from the above date he had four seizures, differing in intensity. On the 29th April, the account given by the father is that his appearance had not improved, that his appetite for meat was irregular, the bowels costive, and the temper more irritable. I then put him up a mixture of nitro-muriatic acid and gentian, and the next report, given on the 25th May, states that the patient's general health had decidedly improved; "his appetite, though still irregular, is certainly better, and he has the appearance of more stamina. He has only had two decided attacks this month, on the 10th, both on the same day, one rather severe one while walking with his mother, and a slighter one two or three hours after. Since that day he has had none . . . so I think we may fairly say he is going on well." The nitromuriatic acid mixture was now changed for a time for decoction of bark and sulphuric acid; but the former was resumed after the 22nd June, because the father wrote that although there had been no return of the seizures, he did not improve as steadily in his general health with the bark. On Sept. 18, I had the satisfaction to receive the following information. "Our little boy has been going on well since I last wrote .... I am happy to say he still continues to improve in general health, and he has had no recurrence of the fits; indeed, we are almost tempted to forget he has ever been attacked, but we still follow out rigidly your rules of diet, early rising, and the sponging bath; his mother says he certainly eats more animal food." I have reason to believe that the boy continues well to the present time.

I shall have to speak of other tonics in discussing the regiminal treatment of epilepsy; but I cannot quit this section without adverting to the use of oleum morrhuæ, which, especially in epileptic children, either by itself or in combination with steel, often proves of the greatest value. The cases in which it is most indicated are those of marked mal-nutrition and presenting a scrofulous diathesis, in which the iodides are also useful, though these must be given cautiously, on account of their depressing influence.

It would be useless to attempt to lay down specific rules for the mode of administering the drugs already spoken of, since the general laws of pathology and therapeutics apply equally to the treatment of epilepsy as to any other disease; therefore, as a matter of course, the endless complications which may accompany epilepsy must be borne in mind, and the necessary remedies ordered accordingly. As long as an irritant of any kind resides in the system it would be next to useless to seek to counteract the spasmodic diathesis; the former must be first removed before we can expect successfully to combat the latter. The weak or diseased condition of any organ, though possibly not bearing any immediate relation to the paroxysmal affection, demands the physician's attention previous to, or in conjunction with, the radical treatment to be adopted. To give a detailed account of all the circumstances that might arise here, would render necessary a review of the whole domain of pathology.

I make these remarks not only as an apology for not entering more fully into the consideration of a host of drugs that may be used in the course of the treatment of epilepsy with more or less advantage; but also as a protest against that specialism, fostered by the public as well as by the profession, which converts every disease into a separate entity, breaks up the unity of the science of medicine, and, in its extremes, does more harm than any extra-professional quackery.

With every disposition and inducement to be brief, however, I think it right to allude to a few further questions in the therapeutics of epilepsy.

There are two classes of poisons which are often found to induce epilepsy, and which are both, in a measure, amenable to the same drug—the iodide of potassium. I refer to the secondary and tertiary forms of syphilis, and to metallic poisoning. It is unnecessary to dwell upon the former, but I wish to say a few words with regard to the latter, as even to the present day the eliminative power of iodide of potassium is not sufficiently appreciated. No metal is more frequently productive of epilepsy than lead, the peculiar depressing and paralysing effects of which are well known. The property of iodide of potassium of liberating and expelling the lead that has been deposited in the tissues of the body was first established by M. Melsens,* and has since been confirmed by various observers.† Where epilepsy results from lead poisoning, the fits with the elimination of the poison, diminish, and may cease altogether, if the intoxication has not operated so long as to generate an invincible habit. During the administration of iodine and the iodides we must be careful not to set up an irritative condition, which

---

* *British and Foreign Medico-Chirurgical Review*, April, 1853.

† Dr. Parkes, *British and Foreign Medico-Chirurgical Review*, April, 1853; On the Detection of Lead in the Urine in Cases of Lead Poisoning, by E. H. Sieveking, M.D. (*Medical Times and Gazette*, 1857); and On some of the Modes of Action of Iodide of Potassium, by E. H. Sieveking, M.D. (*British Medical Journal*, 1857).

would neutralize the therapeutic action of the medicine; nor must we allow it to operate so long upon the kidneys as to impoverish the system, and to induce, as I have known it to do, azoturia.

I shall confine myself to giving a brief summary of one case in illustration of the use of the iodides in epilepsy resulting from the influence of lead.

S. T., a compositor, æt. 20, anæmic, but well built, was sent to me on Feb. 8, 1858, by Mr. Marshall, on account of epilepsy, to which he had been subject for four months, during which time he has had six well-marked seizures. The right hand with which he handles the types has become weak, and inclined to drop; there is rather less sensation in the tips of the right than of the left fingers. He is also subject to pain at the occiput and nape of the neck, supposed to be tendon-pain. The urine high-coloured and scanty. He was ordered to give himself more rest, to use bay-salt sponging, nutritious food, and the following medicines:

> ℞ Iodid. potass. ℨij.
> Syrupi ferri iod. ℨj.
> Aquæ pimentæ ℨv.
> M. Sumat cochleare amplum ter die ex aquâ.
> ℞ Extr. nucis vomicæ gr. ss.
> Extr. aloes aq. gr. j.
> Extr. gent. gr. iv.
> Fiat pilula omni nocte sumenda.

He rapidly improved, the report on Feb. 22 stating that he had no headache, felt better and stronger; good appetite, arm stronger, tongue clean. From this time he took steel and other tonics. No relapse occurred; and on March 16 he is stated to have had no further fits, to have an excellent appetite, and that his affected arm had become as strong as ever. I know that he continued free from epilepsy till April, 1859.

An analogous remedy to the iodide of potassium is the bromide of the same base, which, like the former, is known to possess the power of reducing scrofulous and other tumours, and has been employed with success in some morbid conditions of the uterus. On the occasion of a paper on Epilepsy being read before the Medico-Chirurgical Society,* Sir Charles Locock, the president, remarked that, in epilepsy observing a regularity of return connected with menstruation, he had been led to try the bromide of potassium by an observation made by a German physician that it was capable of producing temporary impotence. Sir Charles Locock stated that he had administered bromide of potassium accordingly in cases of hysteria in which there was a great deal of sexual excitement attended with various distressing symptoms difficult to manage; he found that from five to ten grains, given three times a day, had the effect of calming the excitement to a very marked degree. He stated that fourteen months previously he had been induced to prescribe the remedy in a case of epilepsy connected with sexual excitement after all other medicines had failed; that the result had been an entire cessation of the attacks after they had lasted nine years. Sir Charles added that he had since tried the bromide of potassium in fourteen or fifteen similar cases, and that it had only failed in one. I have since repeatedly prescribed the bromide in epilepsy associated with the symptoms spoken of by Sir

---

* Analysis of Fifty-two Cases of Epilepsy observed by the Author. By E. H. Sieveking, M.D. First Series. (*Medico-Chirurgical Transactions*, 1857).

Charles, and, though I have not enjoyed the same amount of success, have found it decidedly beneficial. In one case, where the irritation of the sexual apparatus was very marked, a permanent cure seemed to be attributable to it.

I have already spoken of the medicines commonly grouped together as anti-spasmodics, such as ammonia, castoreum, assafœtida, valerian, and the like. With the exception of valerian, I hold that none of them exert anything more than a momentary effect upon the system, and that they are useful mainly as temporary stimulants to ward off an attack, which they undoubtedly do under certain circumstances. They are also useful, in combination with other remedies, to relieve symptoms in the intervals of the attacks; but where persistent stimulants are indicated, we may as safely have recourse to the cellar as to the chemist's shop.

Opiates and narcotics have not generally enjoyed a high reputation in the treatment of epilepsy, from the fear of their aggravating the tendency to congestion the brain, and from the frequency with which epilepsy and epileptiform seizures occur during sleep. I have, in an earlier part of the work, detailed a case in which morphia, administered during a severe paroxysm, appeared curative; and I am of opinion that we should do well to employ narcotics more frequently than we do at the commencement of epilepsy, as we can scarcely doubt that during sleep an irregularity in the action of the nervous system supervenes, such as may be met by soothing agents. We should not for that purpose select opium itself, but morphia and its salts, hyoscyamus, conium, belladonna, hydrocyanic acid, and perhaps in some cases chloroform. I have found the inhalation, frequently repeated, of from ten to twenty minims in a child under a year old subject to epileptic convulsions, of very marked benefit. The child was eventually cured; and although I attribute the result to other therapeutic and hygienic agents, I am of opinion that the chloroform aided in diminishing the frequency and duration of the seizures. The effect of the different agents of this class varies materially in its physiological and pathological action, according to which we should determine their respective employment in the disease under consideration. It would be contrary to all knowledge to give opiates in all cases of epilepsy accompanied with costiveness, cerebral congestion, contracted pupil, or belladonna where there is extreme depression, a feeble circulation, and a morbid condition of the sympathetic indicated by dilated pupils. While we should not hesitate to employ these agents in the cases that general pathology points out to be adapted for them, we can scarcely expect them to act in any other way than as means by which we may combat special symptoms. Of late Bretonneau, Trousseau, and others, have specially advocated the employment of belladonna or its alkaloid in epilepsy as a means of radical cure. I have repeatedly found it do harm; but though I have never seen it effect a cure, I have used it with benefit as an adjunct in some cases, especially of nocturnal epilepsy.*

Trousseau gives a pill containing one centigramme of the extract, and an equal quantity of the powder

---

* If I were to generalise on this subject, I should be disposed to attribute the efficacy of belladonna chiefly to its allaying an irritable condition of the organs more immediately under the domain of the sympathetic system. Thus, in other diseases we find it affect especially the sexual organs, allaying irritation and spasm, occurring in their domain.

of belladonna during the first month, by preference in the evening; partly because of the inconvenience attendant upon the remedy at first, and partly because epilepsy is often nocturnal. After a month, he gives two such pills for an equal time, and then three if the remedy is tolerated. If at the end of six or nine months the frequency of the fits is decreased, he knows the disease to be yielding, and he presses his remedy. He states that out of 150 patients he has thus cured 20; but, he asks, will they not relapse?

An agent whose operation is the reverse of narcotic, and manifestly influencing chiefly the sympathetic plexuses, is yet to be specially named, because it has long and deservedly enjoyed a good reputation in the treatment of epilepsy. The drug referred to is digitalis. Its prominent effect upon the economy is shown by reducing the pulse, and by causing a state of wakefulness. It is in the former that we so frequently find an indication for its employment in epilepsy. The pulse is commonly increased in frequency, and of an excitable character, in these cases. In some nothing but the employment of digitalis succeeds in reducing it, and it does so, if properly administered and watched, without any injurious effects. Of course, the patient who takes digitalis must be watched to prevent the toxic effect, the more so as we often find it to be cumulative. This, however, has appeared to me to be less the case in epileptic than in other subjects. I have been surprised at the tolerance of the agent in some patients. Thus, in a young gentleman whose pulse always ranged at from 100 to 120, although his general health was tolerably satisfactory, digitalis was administered in increasing doses for above three months, almost without intermission, before a decided reduction was achieved. I quote the following memoranda of the case from my notes. "Jan. 23, 1860. Has taken three dozen and three bottles of the digitalis mixture (at first Inf. digitalis ʒvii.; tr. ferri muriat. ʒiss., an ounce three times a day, successively ʒi. and ʒiss. of the tincture of digitalis had been added) with scarcely any intermission. Pulse, with one exception, never below 88. Repetatur mistura addendo tinct. digitalis ʒj.

   ℞ Tr. digitalis ʒijss.
    Tr. ferri. mur. ʒj.
    Syrupi rhœados ʒiv.
    Inf. digitalis ad ʒvj.
   Sumat unciam ter die.

Jan. 30. Is very well; the pulse reduced to 76; pergat. Feb. 2. The pulse reduced to 64, and occasionally intermits. Feels very well; omit medicines." The pulse is a safe guide by which to determine the propriety of continuing or stopping this agent. Should sickness supervene, we have a further indication of the necessity of arresting its use. If digitalis exerts a tonic and constricting influence over the vascular system, which those who have watched its effects in uterine hæmorrhage* can scarcely doubt, we may find in that view a reason by which its beneficial effects in epilepsy can be accounted for satisfactorily. Whatever theory we adopt, it can scarcely be denied that in many cases it is at least a valuable adjuvant.

To illustrate how urgently and apparently on good grounds some anti-epileptic remedies are preconised, I may allude to the recommendation of mistle-

---

* See Dr. Dickinson's interesting paper in the *Medico-Chirurgical Transactions* for 1859.

toe by Dr. Fraser. Only so late as 1806, Dr. Fraser* wrote a book to prove that the mistletoe, which then was sinking into oblivion, was a sovereign remedy for epilepsy. He says of it, "If I had not tried it, as far as the limited opportunities of an individual would admit, if I had not found it efficacious in epilepsy, even beyond my most sanguine expectations, I would not now presume to offer it as worthy of the most serious attention of the faculty." "My own experience," he subsequently says, "warrants me in declaring that of eleven cases of epilepsy which were treated with viscus quercinus under my direction in 1802, 1803, 1804, nine were radically cured, one was fatal, and one received no benefit." Dr. Fraser employed the remedy in other cases, but was unable to state with what result, because not a sufficient time had elapsed to pronounce them radical cures. The eleven cases are given in detail, and are not without interest. The following may serve as a sample: A gentleman, aged 23, had been subject to hereditary epilepsy from his third year; after previously taking other medicines without effect, he began the powdered mistletoe, in two-scruple doses, in a draught twice a day. Its administration was continued from the 5th of March, 1802, to the middle of June in the same year, after which the patient remained entirely free from the malady.

The viscus was ordered to be separated from the oak about Christmas, and when dried to be ground to

---

* On Epilepsy and the Use of Viscus Quercinus, or Mistletoe of the Oak, in the Cure of that Disease. By Henry Fraser, M.D. London, 1806. This book is valuable on account of the very complete bibliography it contains bearing on the disease in question.

a fine powder, " which ought to be confined in a bottle, and kept in a situation where both light and air are excluded, as the admission of either tends to deprive this vegetable of its natural efficacy." No less than eighteen different authors are cited by Dr. Fraser as having written specially on the uses of mistletoe in epilepsy, or as having recommended it in the disease.

I have thought it right to give a trial to the mistletoe, but, however it may be suited to promote good feeling and jocularity at our Christmas games, it has not, in my hands, proved to exert the slightest influence over the epileptic paroxysm.

Among the remedies that have of late years attracted a passing attention I may mention indigo and the cotyledon umbilicus. The former was first prescribed by Professor Ideler, of Berlin, and much lauded by him ; it then fell into disuse, because it was thought by physicians employing it that its main effect consisted in colouring the fæces blue. It has recently been again employed by Dr. Rodrigues[*], who relates eleven cases in which he employed indigo in various forms, five of which were cured.

Dr. Rodrigues attributes the want of success which has attended those who have followed Ideler's method to the extreme repugnance excited in the patients by the continuance of the large doses of the remedy. He has therefore modified the plan, and advises the exhibition of large doses for a brief period at the commencement of treatment, so as to make an impression upon the system, and then to continue the remedy at a

---

[*] *Revue Médico-Chirurgicale,* Avril, 1855, and *British and Foreign Medico-Chirurgical Review,* Jan. 1856, p. 250.

reduced rate for a considerable period. The comparative trials made upon different patients lend support to Dr. Rodrigues' views. The indigo may be given in an electuary, in pills or emulsion, and the dose varies from four to fifteen grammes. The following is Professor Ideler's formula :—

    ℞ Indigo gr. xv.
    Pulv. arom. gr. ij.
    Syrupi q. s. ut fiat electuarium.

It is one of the many empirical remedies which we are justified in trying if our rational methods fail us.

Of cotyledon umbilicus,* which has long been a popular remedy, little can be said. I have prescribed it in numerous cases, and generally the patients have appeared to benefit by taking thirty and more grains of the extract three times a day. In one case a cure seemed due to its use; but although the arrest was but temporary, a similar effect has been again obtained on re-administration.† In the paper quoted below another case will be found in which the use of the cotyledon appeared to be followed by a permanent arrest. Others which are there mentioned yielded more doubtful results. The only physiological effect which appears to result from the use of this remedy is increased diuresis; but even that is not very marked.

To review all the individual drugs that have been used and recommended in the treatment of epilepsy would answer no good purpose, and our readers can find very complete lists and full descriptions of their

---

    * A plant belonging to the natural order Crassulaceæ, and growing wild in Dorset and Devonshire.
    † On the Use of Cotyledon Umbilicus in Epilepsy. By E. H. Sieveking, M.D. (*Medical Times and Gazette*, 1854.)

CONCLUSION OF MEDICINAL TREATMENT. 299

properties in the works of Tissot, Fraser, Cooke, Copland, and others.

In fact, there is not a substance in the materia medica, there is scarcely a substance in the world, capable of passing through the gullet of man, that has not at one time or other enjoyed a reputation of being an anti-epileptic.

CHAPTER X.

The hygienic treatment of epilepsy—Influence of air—Exercise —Baths—Special directions for their employment—Quantity and quality of food—Period for taking it—Rest of body and mind—Balance of the mental powers—Education of children —Their home management—Quack-remedies and their influence on the mind—Moral regimen—The physician should be a school-inspector—Conclusion.

I TRUST that nothing I have said in the foregoing chapters will encourage a general scepticism in the value of drugs, properly selected, in the treatment of epilepsy. I hold that in many cases we are able to effect a cure; in most to improve the condition of our patients by their aid. But I am also well assured that important as the proper selection of remedies undoubtedly is, pharmaceutical preparations will fail to secure a satisfactory result, unless we at the same time devote our earnest attention to the hygienic features of the case. I use the term in its widest sense. Nothing that promotes or interferes with the healthy functions of the whole being should be out of the range of the physician's ken. The variety of causes, productive of the malady, bearing more immediately either upon the physical or upon the moral constitution, have been shown to be very numerous. In a disease like epilepsy, where so many and various influences are found to be at work in the production or maintenance of the disorder, where emotions and mental stimuli operate as powerfully as the purely physical causes, we should justly accuse the physician of short-sightedness who neglected these features in combating the malady.

The air the patient breathes, the water he drinks, and his ablutions, his daily occupation and habits, his amusements, the state of his sexual functions, his food and beverages, his clothing, his mental and moral history, his social antecedents, and prospects in life, should be inquired into, in order to determine whether, or in how far, one of these elements may require modification.

The air our patient breathes exercises a most undoubted influence in promoting, if it be impure, that nervous diathesis, that susceptibility of the nervous system, which accompanies or leads to epilepsy. There is no morbid condition to which man is liable, which, if it be not generated, is not aggravated by foul air. In children this is especially visible; and it falls to the lot of medical men frequently to witness the palpably beneficial changes exerted upon an adolescent suffering from epileptiform attacks by improvement in the air that surrounds him. In young children the removal from one room to another often suffices to make a marked impression upon the system. A crowded bedroom, the exhalations of unclean persons or vessels, should be especially avoided, while the bracing and purer air of a country residence is but too frequently an absolute necessity for the recovery or amelioration of the patient. Repeatedly have I seen patients rendered worse by being brought to London from the country, with a view, as is so often done, of obtaining advice when all ordinary remedies have been exhausted. In this case the air doubtless plays an important part; but the excitement of the change, the noise and hurry-skurry of town life, must no less be taken into consideration. On the other hand, I have seen individuals much benefited who enjoyed the quiet and care of a well-regulated home in London, and whose change to the country could confer no other benefit than that of purer and fresher air. Where the means permit of a choice of localities, regard must be had to the constitution and fibre of the individual, and the general rules regulating the choice of climate apply as to other diseases. The same remark holds good with regard to British and foreign watering-places. The views we adopt with regard to the phases shown in the pathology of the disease, will guide us in making the selection.

The value of exercise properly conducted can scarcely be overrated, but it should be employed with a special reference to obtaining a fuller supply of the oxygenating medium. Walking on the flags of London, with all the din of a noisy populace around, irritating every fibre in a delicate constitution, is not the exercise likely to benefit; while the exhilarating and beneficial influence of a walk through the fields and lanes of a rural district is undeniable. The amount of exercise to be taken must depend upon the habit and strength of the individual; it should never be carried to the extent of exhaustion, and be of a kind to place the patient in positions of danger. Riding is for this reason objectionable. One of my patients broke his arm in consequence of being seized with a paroxysm while on horseback.

Travelling, both by sea and land, may, under certain limitations, be advisable, both on account of the healthy stimulus it gives to the mental functions, and on account of the improvement in the nutritive sphere which it generally induces.

Next in importance to the air in the hygienic treatment of epilepsy or its congeners, is the use of water, as beverage undoubtedly, but still more as a roborant, externally applied. To those who are habituated to the daily use of the shower or sponge bath, it seems almost impossible to exist without them. And yet even with us, proverbially a cleanly people, it is surprising how many go, from the beginning to the end of the year, in utter ignorance of the purifying and invigorating influences of a general bath. The discomfort resulting from the omission of the daily bath, the feeling of restlessness and almost feverishness which affect us, when from accidental causes it has been passed over, are feeble indications of the derangements of the nervous system which must ensue when the ablutions, for months and years together, are confined to the face and hands. As an hygienic application, then, the daily use of the cold bath on rising is to be ordered, where there are no special grounds that counter-indicate it. I would not undertake the treatment of a case of epilepsy in which its use, advised by myself, were objected to. The period of using it, as well as the temperature, deserves consideration, and still more the length of time during which the patient is to remain in the water. The more feeble the patient the more it will be necessary to elevate the temperature, which should be gradually lowered until he is able to bear it at the temperature of the air. If the feebleness be great, the bath should not be used until after breakfast, and it may even be advisable that this meal be taken before the toilet is arranged, in order that the patient may not have to go through the ordeal of dressing twice over. The invigorating influence of the cold sponge-bath may be much enhanced by the addition of two or three pounds of bay-salt, in which form it is one of the most admirable tonics we possess; it may, however, be well to remember, that to some persons with a delicate skin the bay-salt proves too irritating, and causes prurigo of so severe a character as to render its omission necessary.

The shower-bath is a mode of applying cold water to the surface which should not be used as indiscriminately as it is; the shock is too great in many instances, and the reaction by no means certain to be sufficient.* A

---

* The depressing influence which the cold shower-bath exercises upon the circulation in health is very marked. Some years ago I instituted some experiments upon myself with reference to this point, of which the following is a brief summary:—

They were divided into two series; the first consisted of twenty observations, each comprising three examinations of the pulse; it was counted soon after rising; it was again counted after taking a uniform amount of dumb-bell exercise, and again soon after the shower-bath, the precaution, recommended by Dr. Graves, being observed in each instance of allowing about a minute to elapse after the occurrence, the influence of which upon the pulse it was desired to estimate. The averages obtained in this series were—

Pulse on rising.	Pulse after exercise.	Pulse after shower-bath.
69·50	76·90	68·85

The average increase of the pulse, therefore, caused by the exercise was 7·40; the average depression caused by the bath was 8·05.

In the second series of thirty observations the bath was taken immediately after rising, and the exercise followed the bath. The averages here were—

Pulse on rising.	Pulse after shower-bath.	Pulse after exercise.
69·17	63·03	66·24

The average depression caused by the bath was 6·14; which

certain amount of physical vigour is necessary to justify its employment; and it will always be necessary carefully to watch its effects in disease, in order that we may not neutralize our good intentions by the excess of the remedy. The surface reaction which follows the cold bath will always be a safe indication of its appropriateness; if a feeling of comfort and warmth does not ensue, if the patient feels overfatigued, instead of refreshed, when it is over, he will not be benefited. The reaction should, however, always be promoted by vigorous friction of the whole surface with rough towels, Baden towels, horse-hair rubbers, and the like; and this should not be left to the patient, if we are not certain that his strength is ample for the purpose. The passive movement induced by friction is in itself tonic and soothing, and should not be omitted even when, for temporary reasons, the bath is counter-indicated. With regard to the temperature of the water, we must remember that the feebler the patient the more we should elevate the temperature of his bath. A bath that may be tonic to one person at 80° would be anything but invigorating and refreshing to a person whose stamina justified its reduction to 70° or 60°. A shower-bath will be tonic at a higher temperature than a sponge-bath, owing to the rapid cooling of the water as it descends, and also to the shock which the fall produces.

The warm-bath can scarcely be regarded as an hygienic measure; its employment depends upon definite indications; and these cannot be determined

---

the dumb-bell exercise was not able to remove, the average increase of the pulse caused by it being only 3·21. It follows that the reducing effect of the bath is considerably greater than the exciting effect of the exercise; or, given in numbers, as 6·14 to 3·21.

by other than the medical man. Its habitual use by a healthy person is enervating; and though it may be occasionally suited to the epileptic patient, it is not adapted for daily employment.

The cases in which I employ warm baths are those in which there is an unusual excitability of the whole system. Where with tonics soothing remedies are indicated, the introduction of salts of iron* into the warm bath very suitably assists in meeting both indications. I have repeatedly derived great benefit from the administration of a series of warm baths (at 96° F.) medicated in this way. The precautions to be observed in the use of the warm bath are, that it should be given at the close of the day, that it should not be taken on a full stomach, that the patient should avoid the possibility of taking cold after it; and, therefore, as a rule, retire to bed as soon as possible.

In discussing the dietary of an epileptic we must pay careful attention to three points, each of which even a limited experience in the disease will show to possess an important bearing; they are the quality of the food, its quantity, and the period of consumption.

The food of the epileptic patient, as a general rule, must be nutritious and copious. Among the lower orders and labouring classes poverty often forces the patient to be content with inadequate and unwholesome diet, and is thus an accessory cause to epilepsy, promoting and fostering the atony of the vascular and nervous system, which in the majority of cases is at the bottom of the whole malady.

---

* Mr. Twinberrow, of Edward Street, Portman Square, has at my suggestion prepared cakes of potassio-tartrate of iron, gum, and honey, each containing one ounce of the chalybeate, which are convenient for this purpose.

Among the higher classes we are more likely to have to deal with repletion as an exciting cause; in both, indigestible food is only too often and justly accused as being the immediate inducement of a paroxysm. Milk or cocoa, soft-boiled eggs, stale bread, or toast, broiled bacon, or a mutton chop, are suitable for breakfast; and in many cases a glass of milk and a biscuit taken before the morning bath are a good introduction to the day's work. The chief meal of the day, which should be taken at one o'clock, should consist of well-cooked white fish, mutton or beef, fowls or game; pork and veal, made-dishes of all kinds, being proscribed for all ages, but especially the first-named articles. Among vegetables, well-boiled potatoes, carrots, turnips, cauliflowers, artichokes, are permissible, while broad beans, peas, fresh or dry, greens, and all things found to induce flatulency, are to be avoided. Farinaceous, light puddings of bread, sago, rice, arrowroot, macaroni, are to be commended; while pastry of all kinds, creams, trifles, and the usual course of party-sweets, are to be deprecated. Ripe fruit is not in itself to be disapproved of, but, as a rule, it is best avoided after a previous meal. I have never found strawberries, raspberries, grapes, cherries, oranges, do harm, nor do I see any objection to stewed fruits generally, excepting, perhaps, plums; I except French plums for the same reason that I should object to figs, currants, and raisins, that their thick skins, which are always eaten, are extremely indigestible, and are frequently a source of dyspeptic derangement. Among the fruits to be especially proscribed, nuts of all kinds deserve to be singled out.

There should be a meal at 5 or 6 P.M., to consist of milk or cocoa, and bread and butter, and a light supper would be requisite for an adult an hour or two before retiring to rest, which should consist exclusively of milk and farinaceous materials. If it is unavoidable that the chief meal of the day is transferred to a late hour, it need scarcely be said that a supper of any kind is highly objectionable.

The employment of wine or beer must depend upon circumstances. Anything approaching to intemperance in the epileptic should be guarded against, even more than in other persons; but even the temperate use of alcoholic beverages is to be advised only with great caution. Where the tendency to cephalic congestion is marked, they commonly increase the headache and flushing, and are therefore counter-indicated; this is especially the case with regard to taking these beverages shortly before going to bed. Where the patient does not suffer from headache, there is a pale, chilly surface, and other evidences of generally defective innervation, the moderate use of wholesome beer or wine will prove beneficial. As a general rule, tea and coffee had better be exchanged for milk and cocoa for the morning and evening meals. We constantly meet with instances in which the prolonged abstinence from food, accompanied by hard work and more or less mental anxiety, produces the epileptic paroxysm; and where by ordering food at proper intervals the patient is able to go through his work without inconvenience. I am frequently in the habit of meeting a gentleman who is epileptic, but who, since I have advised a substantial repast in the middle of the day, has had no return of his attacks. Previously he walked a considerable distance to and from his place of business, at which he was occupied from an early morning hour till late in the afternoon. In this case it was not the fatigue, but the inadequacy of the machinery to support the fatigue, which induced the paroxysm.

Many other cases might be cited in illustration of the previous remarks, but no medical man will read them without recalling instances from his own practice.

I would make one remark in regard to suppers. They are, now-a-days, tabooed, because the fashion has thrown the dinner hour so late that the dinner has itself become a supper. But for those who still indulge in early hours, an early dinner, with a five o'clock cup of tea and thin slice of bread and butter, does not suffice for the sustenance of the body to the performance of its proper functions. The sensation of hunger does not express itself with the same urgency as that of thirst. We often find delicate, nervous subjects in a state approaching to starvation because they do not rightly interpret the languor and debility that creeps over them, and which is in reality hunger. The stomach may undoubtedly be trained to a state of inability, to do less than its proper amount of work; in such cases the introduction of food more frequently, and in increasing quantities, becomes a *sine quâ non* in the restoration of health. Here a light supper may prevent the inanition which has caused the fits, and arrest them; while the exhibition of pepsine and similar preparations may enable the stomach to recover its tone and vigour.

The quality of the food should always be nutritious and digestible, well-cooked meat forming the staple article; all hard, greasy, sour, indigestible substances require to be specially prohibited from the dietary of the epileptic. In children, too, and old people, the mastication should be watched over, that the stomach be not charged with a "rudis indigestaque moles" of unchewed meat and vegetables.

In many cases we find that no remedial agents or hygienic suggestions will enable our patients to go through their ordinary occupations without a frequent occurrence of paroxysms, until we order them complete rest. Rest of body and mind, as we all know, is one of the most certain restoratives to a healthy state; but nowhere is its beneficial influence more palpable than where disease of an exhausting character has fastened upon the nervous system. Again and again have I had occasion to tell my patients that there was no hope for an alleviation of their malady until they could abstain from their ordinary occupations; and though mere abstinence from work does not cure epilepsy, I have many times satisfied myself that it may cause a temporary arrest of the fits, which by proper treatment may be converted into a cure.

Under the term rest are to be comprised the avoidance of all undue stimuli to the body, and yet more to the mind, which the conventionalisms of society so frequently entail upon us, and which become disproportionately irritating the more susceptible the individual. The harassing cares of a large family; the attendance upon a sick friend, relative, or superior; visiting crowded assemblies at all times and of all kinds, especially at a time when the patient ought to have given himself up to Morpheus; over-straining the mind in the forcing-house system adopted in many schools; are a few of the many noxious influences to be sedulously guarded against or removed in the epileptic.

And here I would take the opportunity of again inculcating, as has so often been done, but never can be done sufficiently, the necessity in early education of attending to the preservation of the due balance between the bodily and mental powers. While the processes of physical nutrition are at their height, while new impressions are constantly acting upon the susceptible nervous system, every undue tax upon the mind is laid under a heavy penalty, which physiology or nature will certainly enforce in some shape or another. The children who promise best, whose organization seems endowed with higher capabilities than the average, are the very children who require most watching; for in them the danger of the body being sacrificed is greatest, and the result will be incapability, at man's age, of executing those duties which may be expected of a vigorous adult. The body is the mind's agent, but if the machinery is creaky it cannot fulfil its legitimate objects. The pupil-teachers of our national schools are a class of children among whom many of these over-wrought brains are found, and are often brought to the hospital on account of cephalic affections, manifestly the result of an attempt to force the mind into maturity without reference to the vigour of the body.

But we find the same mistakes committed in all ranks of society, and it cannot be put with sufficient force that the object of education is to prepare the child for the hard work of real life, to steel him in body and mind for the harassing cares and buffets that he must support when he comes to adult life, unless he is to sink under them, a prey to illness and anxiety. All knowledge or accomplishments that do not promote the development of the mental faculties in such a way as to be available and bring fruit in after life, that choke rather than oxygenate and vivify, are so much poison.

Parents are often quite as much to blame in the home management of children as schoolmasters. The trials of temper to which they are subjected by fractious irritable children are great; still, there is commonly a mutual fault, which the candid parent will on self-examination be ready to admit.

A whole chapter on the moral management of the nursery might not be out of place here, did it not occupy more space than I can devote to the subject. I limit myself to indicating the necessity of attending to these matters, of seeking to train the child by precept and example in habits of occupation, self-control, and forbearance to others, which are scarcely to be acquired in after life, or only at the sacrifice of means and time that ought to have been devoted to other purposes. The application of the principle in its details must be left to the individual parent or tutor.

There is often a difficulty in ascertaining with sufficient accuracy the various domestic influences to which a patient is subjected. But it is quite certain that the physician will not be informed if it is not a matter upon which he lays stress, as the majority of patients consider it sufficient to communicate their subjective sensations to the medical man, expecting that for each symptom he has an appropriate remedy. That all medical men do not attribute sufficient importance to hygienic measures is but too patent. As an instance of the variety of injurious influences to which so many of our patients are exposed before or during the time that the epileptic fits are occurring, I quote the following graphic account given by a patient, in the north of England, who consulted me by letter. I premise that the patient was first seized with epilepsy, no essential symptom of which was wanting to establish the diagnosis, two months after confinement with her first child :—

"That you may judge better as to the cause of my anxiety, and the labour I have gone through, and the excitement necessarily attending, I will state that after my marriage we lived in the country at ———. We had a few acres of garden-land. My husband, being a tanner, worked at his trade, and as I was always of an active turn and wanted to make all I could, I used to work a good deal in the garden, and also attended many markets—Sheffield, Barnsley, and Wakefield; to the three latter I had to travel in the night, with a horse and cart, leaving home about half-past eleven at night; this exposed me much to the night air. This was before I had the fits, and for a few years after, and at the time Mr. ——— attended me." Circumstances caused removal from the country to town, where the parties kept a shop: and the writer states that, from her husband "not being a scholar," she had all the business matters to attend to. In addition to these necessary claims upon her, she says, "I have read and studied a great deal, and often when I ought to have been in bed. . . . . I have also been very fond of singing; before my marriage and long after, I used to sing in the choir at chapel; but I have not done so for a long time, except on a few special occasions. I have often been told this excited me too much, but as I never had a fit immediately after excitement, I never could think it was the cause of them; but I cannot stand any of these things as I once did."

This extract indicates at once the various influences to which, more or less modified, our epileptic patients are commonly subjected, as well as the mode of argument adopted by them, and often, too, by medical men. The fits did not immediately follow upon one of the occurrences or occupations adverted to, therefore there was no causative relation between the two. Excessive bodily fatigue, then, and excessive mental exertion and excitement—and, yet more, a combination of the two—are to be guarded against wherever there is a tendency to spasmodic action, and, *à fortiori*, where the epileptic paroxysm has already occurred, even more than in a healthy individual. The younger the individual, the greater the necessity for these precautions.

It will not, however, suffice to prohibit only: as a good government seeks to supersede the terrors of the law by improving schools and establishing reformatories, and endeavours to substitute premiums for good conduct, for the degrading influence of fear; so should the physician facilitate the patient's abandonment of prejudicial habits by suggesting modes of relaxation and occupation for those which he deprives him of. For the morbid stimulus he must substitute healthy food, whether for mind or for body. And he must not take for granted that, when he prohibits one thing, the patient will necessarily select the thing that is right. Such careful directions are as necessary as it is that an officer who, intending his troops to leave certain quarters, and ordering them not to march by a given route, should command the road that he requires them to take. Were he to issue the former order without the latter, his soldiers might be as much and more at a loss than if he had given none at all. Though we wish to obtain rest for our patient, that does not mean sloth either of mind or of body; hence such amusements and such engagements must be sought out as will meet his peculiar capacity and requirements. Here the medical man will fail without an intimate knowledge of his client's habits and character; but without it he can neither be truly successful, nor will he gain his patient's confidence.

The confidence of the patient is so important an element in the treatment of diseases of the nervous system, that no one can be considered suitable to treat them who does not possess the happy talent of inspiring it. The influence of mental conditions in arresting epilepsy is shown in the fact, pointed to by Esquirol and others, of the temporary benefit obtained in epileptics by the exhibition of any or every new remedy that was believed by the patient to promise a cure. While I willingly admit that the fleeting amelioration obtained was evidence of the remedy not in itself possessing curative powers—at least, in the way in which it was administered—I would regard the benefit that resulted as an encouragement not to neglect the moral element.

As fright and many emotional influences have been shown undoubtedly to produce the paroxysm, so, with the other evidence before us, we are justified in believing, that influences that exalt and invigorate the will, can allay the storm. The history of the therapeutics of epilepsy brings down to us lists of remedies so unpleasant—nay, so abhorrent—that it is impossible to conceive that they should ever have been employed, had not a strong conviction on the mind of the patient, that they would benefit him, occasionally realized his hopes. Pliny, for instance, records many such; he objects to some of the specifics for epilepsy that were recommended in his time; of others he approves, such as eating the testicles of a ram, gall mixed with honey, &c.; but, wherever he mentions a very peculiar remedy, he throws the responsibility on the shoulders of the Magi; thus—

"Comitialibus detur et lactis equini potus, lichenque in aceto mulso bibendus. Dantur et carnes caprinæ in rogo hominis tostæ, ut volunt Magi."* And again: "Magis placet draconis cauda in pelle dorcadis alligata cervinis nervis, vel lapilli e ventre pullorum hirundinum sinistro lacerto annexi." Forestus, as we are told by Dr. Cooke,† administered an arcanum, handed down to him by Guainerus, consisting of human cranium and hoofs of an ass in powder. Tissot, who in this as in everything else connected with epilepsy supplies us with a mine of information, enumerates many substances which were reputed as anti-epileptic remedies. Among the chief of those which he condemns as useless are, earthworms taken on an empty stomach before sunrise in June, or at the moment of coitus; the foot of an elk; of a hare; the afterbirth of a firstborn child; powdered human skull which had not been buried; scrapings of the vertebræ of a man killed by violence; human brain; crow's brain; and the like.

The great objection, as Dr. Cooke wisely remarks, in speaking of employing such and analogous means as those just alluded to, and which manifestly operate only by the influence they produce upon the mind of the patient, is, that we are unable to measure their effect. Their employment may, and often has, produced irreparable mischief; as in the case of a young man, related by Tulpius, who took a draught of human blood with great repugnance, and at once became much worse: "Tantum abest," he says graphically, though not in

---

* Plinii Secundi Hist. Natur. Tomus tertius, lib. xxviii. cap. xvii.; and lib. xxx. c. x. Roterodami, 1668.
† A Treatise on Nervous Diseases. By John Cooke, M.D. Vol. ii. London, 1823.

Ciceronian style, "ut terribilis morbus inde imminuerctur ut potius plurimum incrementi sumpserit, habueritque multo pejus quam ante Thyesteam hanc mensam."
While we should scout all means of acting upon the patient's *morale* which are not consonant with our knowledge of psychology and with high-toned morality,*

---

* In connexion with this subject I would direct the reader's attention to an admirable article on French and German Psychology by Dr. Sheppard in the *Brit. and For. Med.-Chir. Review*, for October, 1860. I cannot refrain from quoting the following passage which appears apposite to the previous remarks: "There are incidents never suspected or alluded to, fraught with overwhelming power, and capable of effecting for all time, and something beyond time, the destinies of an immortal being. A silly name given by silly parents, or a pair of bandy legs given by Nature, have been the absolute ruin of many a boy of nervous and impressionable temperament, and surrounded him with danger from the cradle to the grave. As thus, the name and the legs have provoked habitual laughter—that weapon, as Luther expresses it, which 'disconcerts the devil and makes him run like a fool'—the laughter has influenced character, the character has influenced life, and life and death together have influenced, oh, what tremendous interests! In this way, a physical deformity conditions a specialty for its unhappy victim, which may embarrass at every moment of his career and every act of his life. And a baptismal appellation, fit rather for the reign of Jehoiakim than that of Victoria, though prompted by parental piety, may eventuate in filial alienation. We have known a little Habakkuk at a royal foundation who was literally crushed under his prophetic nomenclature, and who, teased and tormented, died delirious; whereas, had he been John or William, he might yet have been living."

These questions, with the whole subject of bullying at schools, public and private, are receiving much more attention now than formerly; see the literature of the recent Revivals, the "Tom Brown" literature, including of course the prefatory remarks to the later editions of *the* "Schooldays," and other works. But we we may not lose sight of mental influences calculated to soothe and invigorate the irritable and feeble, to rouse and stimulate the torpid and indolent. The former class will most frequently come under our notice in connexion with the disease under consideration.

We shall not be able effectually to aid them without an intimate knowledge of the domestic relations, habits, and occupations of our patients. The persons that surround the patient, the books they read, the studies and pursuits they are engaged in, the amusements they are devoted to, the hobbies they ride, are all subjects which the physician must not consider beneath his attention if he wishes to be anything more than a mere prescriber.

It would be impossible to lay down rules of treatment applicable to individual cases. The mode of procedure here, as in a physical point of view, must vary with the individual to be subjected to it. Those who admit and adopt the principles contended for, will, with comparative facility, adapt them to each case. But whatever moral *régime* be adopted, it must always

---

shall never find that society at large will perform her duty to the young generation until its teachers quit all idealism, and learn to appreciate physiological truth, and its bearings upon the healthy training of mind and body. Here is one of the greatest spheres of the physician, which, however, he rarely has the means of adequately filling until he has consumed all his energies in the more urgent necessity of providing for the daily wants of his family. Could not some of the inspectorships of schools be given to enlightened members of the medical profession, than whom none can be found more competent to fulfil the highest duties of these offices, even including the supervision of the manner in which the various branches of technical knowledge are taught?

be selected and carried out so as to avoid violence to the patient's feelings, and not to impair his confidence in his adviser.

A few words on the points suggested will conclude what I wish to say on the subject. It often happens that, in domestic circles, a feeling of irritation may be excited and kept up even unintentionally by want of regard to the wishes, or, on the other hand, by extreme consideration for the patient.

We all know instances of domestic discomfort and unhappiness where we have no reason to doubt the existence of real mutual affection. In numerous families especially, the delicate child is often subjected to annoyances by the more robust, which can scarcely be obviated or prevented.

It is often found that removal into other scenes, and association with different people, will cause the child or the adult to exert an amount of self-control that the same person appeared incapable of at home. With the greater self-command, the irritability and susceptibility of the nervous system will also be toned down, and greater vigour will be substituted. For another child, the removal from school or from associations of a vicious character will be of paramount importance.

With children, as with older persons, the character of the mental stimuli provided in the shape of books and other occupations, in the times not devoted to study, to a profession, or to a business, merits the consideration of the physician. Frivolous reading, books that exclusively excite the emotions, though in themselves of no immoral tendency; but, above all, books that are calculated to give an improper stimulus to the sexual feelings,—from "Lemprière," the bane of schoolboys, to "Don Juan" or "Les Mystères de Paris," the poison of unripe adolescents—should be sedulously avoided. For similar reasons, games of hazard, concerts, theatres, and balls, which should no less be prohibited on sanitary grounds, are to be specially inhibited for the young people under consideration. Those abortions of modern civilization, full-dressed children's parties, which are got up for the sake of pandering to the love of display and vanity of parents much more than for the amusement of the children, must also be mentioned as things specially to be avoided where the nervous system shows symptoms of unusual sensibility or weakness.

How much of the restlessness of mind that is the source of unhappiness to the adult, has its foundation in early home culture, or rather in its neglect! Instead of training the child to look to itself and the resources that are within its reach for amusement and occupation, parents are constantly holding out some great stimulus as an inducement for the performance of its duties, while the best motives are put aside or not encouraged. The life of a child in large towns has many drawbacks which a country child does not experience; but most children have a definite taste which only requires to be brought out to give them a field for healthy activity. Drawing, reading aloud, carpentering, singing, botanizing, drilling, are some of the means by which spare hours may be most pleasantly and wholesomely filled. But, then, it is necessary that there be in the parents a capacity to see the child's bent, and give it a proper direction. I must, however, check my pen lest I should wander from the path that I have sought to confine myself to through these pages.

I should exceed the limits that I have proposed to this inquiry were I to go more into detail.

I may sum up the remarks on the hygienic treatment of epilepsy and allied diseases—a subject that demands on the part of the physician an intimate appreciation of character, and of the mutual influences of the psychical and physical functions—with the caution, to examine every case on its own merits, to decide upon the treatment to be adopted according to the conclusions thus arrived at, and to consider no matter which may have a bearing upon the social, or moral, or physiological circumstances of the patient as beneath the dignity of science.

APPENDIX OF FORMULÆ.

I SUBJOIN a few of the formulæ which I have been in the habit of employing, the particular indications for which will appear from the foregoing pages. In practice, many variations and combinations are necessary which the medical man will determine upon according to the particular features that present themselves. Still it may be convenient to the junior practitioner to be able to refer to prescriptions that have been actually employed, and that have proved more or less satisfactory in the writer's hands. It is scarcely necessary to add that they might have been multiplied indefinitely, if the intention had been, even remotely, to exhaust this part of the subject.

Mistura Potassii Bromidi.

℞ Bromidi potassii ℈viij.
Tr. zinziberis ʒiv.
Inf. lupuli ℥vijss.
M.—S. Cochlearia dua ampla ter die sumenda.

Mistura Peptica.

℞ Pepsinæ ʒiij.
Acidi hydrochlorici ♏xlviij.
Aquæ pimentæ ℥vij.
M.—S. Cochleare amplum ter die ante cibum sumendum.

Mistura Terebinthinæ.

℞ Olei terebinthinæ ʒj.
 Misturæ amygdalæ ʒviij.
M.—S. Cochlearia duo ampla ter die capienda.

Mistura Zinci Sulphatis.

℞ Zinci sulphatis gr. ij.—gr. xviij.
 Inf. valerianæ ʒss.
M.—Sumat æger cochleare amplum ter die.

Mistura Valerianatis Zinci.

℞ Zinci valerianatis Ɉiv.
 Acidi lactici ʒij.
 Inf. lupuli ʒviij.
M.—S. Cochlearia dua ampla ter die capienda.

Mistura Acida Tonica.

℞ Acidi nitro-hydrochlorici ♏xl.
 Syrupi aurant. ʒj.
 Inf. aurant. co. ⎫
 vel Inf. cascarillæ ⎬ ʒvij.
 vel Inf. gent. co. ⎭
M.—S. Cochlearia dua ampla ter die sumenda.

Mistura Tonica Stimulans.

℞ Ferri et quinæ citratis ʒj.
 Syrupi zinziberis ʒvj.
 Sp. etherchlorici ʒij.
 Aquæ cinnam. ad ʒviij.
M.—S. Cochleare magnum ter die sumendum.

Mistura Strychniæ.

℞ Strychniæ gr. j.
 Acid. sulph. dil. ʒj.
 Tr. aurant.
 Syr. zinzib. āā. ʒvj.
 Inf. gent. co. ad ʒviij.
M.—S. Cochleare amplum ter die sumendum.

Mistura Digitalis Comp.

℞ Tr. Ferri mur. ʒj.
Tr. digitalis ʒiss., ʒiiss.
Syrupi rhœados ʒiv.
Inf. digitalis ʒvj.
M.—S. Cochlearia dua ampla ter die sumenda.

Mistura Ferri et Strychniæ.

℞ Ferri et strychniæ citratis gr. xxxij. ad gr. lxiv.
Syrupi ʒj.
Aquæ ʒiij.
M.—S. Cochlearia parva dua ter die sumenda.

Pilulæ Camphoræ Co.

℞ Camphoræ gr. xv.
Spir. vini q. s.
Extr. hyoscyami gr. xv.
M.—Fiat massa in pilulas vj. dividenda.
S. Pilulæ binæ omni nocte sumendæ.

Pilulæ Zinci Oxidi.

℞ Oxidi zinci.
Extr. gentianæ, vel extr. lupuli āā
Fiant pilulæ granorum quatuor.
Sumat æger pilulam ter die, et augeatur dosis quartâ quâque die unâ pilulâ.

Pilulæ Belladonnæ.

℞ Extracti belladonnæ gr. iss.
Extr. gentianæ, vel pilul. conii. co., vel conf. arom. Ɔj.
M.—Fiat massa in pilulas vj. dividenda.
Sumat æger pilulam omni nocte (et si necesse sit augeatur dosis).

Pilulæ Zinci Valerianatis.

℞ Zinci valerian gr. ij.
Extr. gentianæ q. s.
M.—Fiant pilulæ tales xlviij.
S. Pilula ter die sumenda, augeatur dosis quartâ quâque die.

Pilulæ Belladonnæ Co.

℞ Extr. belladonnæ gr. ¼
Extr. coloc. co.
Extr. gentian. āā gr. ij.
M.—Sumat pilulam ter die. Mitt. tales xviij.

Pilulæ Nucis Vomicæ.

℞ Extr. nucis vomicæ, gr. iij.
Pil. aloes co. gr. xxiv.
M.—Divide massam in pilulas vj.
S. Pilula horæ somni sumenda.

Pilulæ Valerianæ Co.

℞ Ferri valerianatis
Zinci valerianatis āā gr. xij.
Extr. gentianæ gr. xxiv.
M.—Divide massam in pilulas xij.
S. Pilula ter die sumenda.

Pilulæ Ferri Sulphatis Co.

℞ Ferri sulphatis gr. xij.
Extr. rhei gr. vj.
Extr. gentianæ gr. xlij.
Ol. cinnam. ℳ vj.
M.—Ft. massa in pilulas xij. dividenda.
S. Pilula ter die sumenda.

FORMULÆ.

Pilulæ Zinci Lactatis.
R Zinci lactatis ʒj.
Confect. rosæ ʒij.
M.—Divide massam in pilulas lx.
S. Pilula ter die sumenda—tertiâ quâque die augeatur dosis, ut jussum est.

Pilulæ Ferri et Quinæ.
R Sulph. ferri
Sulph. quinæ, āā gr. xxív.
Ext. gentian. ʒiss.
Ol. absinth. ℥xv.
M.—Divide massam in pilulas xxiv.
S. Pilula ter die sumenda.

Pilulæ Zinci et Digitalis.
R Sulph. zinc. gr. xxxvj.
Pulv. digitalis gr. iij.
Extr. rhei gr. viij.
Extr. lupuli ʒj.
M.—Divide massam in pilulas xxiv.
S. Pilula ter die sumenda.

Pilulæ Zinci Sulphatis.
R Zinci sulphatis ʒj.
Extr. rhei gr. xij.
Extr. gentian. ʒij.
M.—Divide massam in pilulas lx.
S. Pilula i.—xv. ter die sumenda vel amplius.

Pilulæ Ferri et Belladonnæ.
R Ext. belladonnæ gr. ij.
Ferri sulphatis gr. xxiv.
Ext. lupuli ʒj.
Ol. cinnam. ℥vj.
M.—Divide massam in pilulas xxiv.
S Pilulæ binæ bis die sumendæ.

Pulvis Pepticus cum Strychniæ.

℞ Pepsinæ gr. xv.
   Strychniæ gr. $\frac{1}{24}$
M.—Ft. pulveres tales viij.
S. Pulvis ter die cum cibo sumendus.

Pulvis Zinci Phosphatis Co.

℞ Zinci phosphatis gr. x.
   Strychniæ gr. $\frac{1}{24}$
M.—S. Pulvis ter die sumendus.

Pulvis Sulphuris Compositus.

℞ Flor. sulphuris
   Magnesiæ carbonatis āā gr. x.
   Pulv. rhei gr. v.
M.—Fiat pulvis horâ somni sumendus.
    (For an adult).

Guttæ Antispasmodicæ.

℞ Spir. ammon. arom. ʒiv.
   Spir. ether. chlor. ʒij.
   Tr. castorei ʒiij.
   Aq. flor. aurant. ʒvij.
M.—S. Cochleare parvum pro re nata ex aquâ sumendum.

℞ Spir. ammon. arom.
   Tr. castorei
   Tr. lavand. co.
   Syrupi zinzib. āā ʒiv.
M.—S. Cochleare parvum sæpius quando convulsiones urgeant sumendum.

Pulvis Phosphori.

℞ Phosphori amorphi gr. iv.
   Ferri carb. cum saccharo gr. iij.
M.—F. pulvis. Mitte xviij.
Sumat pulverem ter die.

Pulvis Pepticus Co.

℞ Pepsinæ gr. xv.
Ferri carb. c. sacch. gr. x.
M.—S. Pulvis ter die cum cibo sumendus.

Electuarium Zinci Oxidi.

℞ Zinci oxidi gr. clx.
Sacchari fæcis ʒij.
M.—S. Cochleare parvum ter die sumendum.

Lotio Sedativa.

℞ Morphiæ acet. gr. xij.
Acid. hydrocy. dil. ʒiv.
Aquæ rosarum ʒxiss.
M.—Lotio mane et nocte partibus affectis applicanda.

Lotio Refrigerans.

℞ Muriatis ammoniæ ʒiv.
Nitratis potassæ ʒij.
Spirit. vini ʒij.
Aquæ font. vel aquæ rosæ ʒx.
M.—Fiat lotio, capiti pro re nata applicanda.

Vel

℞ Muriatis ammoniæ ʒiij.
Aceti,
Spir. vini āā ʒiij.
M.—Lotio, tribus aquæ partibus admixta, applicanda.

INDEX.

ABYSSINIANS, sorcery among, 88
Accidents in fit, 4
Acids, mineral, 285
  ,, nitro-muriatic, 235
ADAMS, quoted, 73
Air, influence of 301
Albuminuria, 116
ALISON, case of Dr., 59
Allarton's steel biscuits, 272
ANDREE, quoted, 21
Antimony, as a counter-irritant, 261
Antiphlogistic treatment, 270
Apoplexy, its distinction, 29
  ,, its relation to epilepsy, 114
Argyria, case of, 284
Army statistics, 96
Asylum Journal, remarks in, 62
Aura, 4, 14 et seqq.
  ,, cerebral or non-cerebral, 23
  ,, its value as to treatment, 23
  ,, its artificial arrest, 240
  ,, its character and varieties, 14 et seqq.
  ,, various forms of, 22

BALFOUR, quoted, 96
Baths, cold, 303
  ,, warm, 305
Bay salt, 304

Beer, its use, 308
BELL, quoted, 103
Belladonna, 293
Bloodvessels in medulla oblongata, 181
BLOOMFIELD, on demoniacs, 89
BLOT, on albuminuria, 117
BOERHAAVE, quoted, 28, 30
BOILEAU, quoted, 266
BOSTOCK, on uræmia, 116
BOUDIN, quoted, 100
Bowels, state of, 122
BOYD, quoted, 48, 164, 185
Brain, weight of, in epilepsy, 166
  ,, congestion of, 222
BRETONNEAU, quoted, 293
BRIGHT, quoted, 28
  ,, on albuminuria, 116
BRODHURST, quoted, 74
Bromide of potassium, 291
BROWN-SÉQUARD'S experiments, 201
BUCKNILL, quoted, 86
BURNETT, quoted, 41
BURROWS, on cerebral circulation, 153

CALMEIL, quoted, 145
CARPENTER, on sleep, 153
  ,, on epilepsy, 196
Carotids, ligature of, 196
  ,, compression of, 236

INDEX.

Causes of epilepsy, 85 et seqq.
„ exciting, their difference, 144
„ their mode of operation 144
„ relation of predisposing and exciting, 157
„ psychical and physical, 145
Cautery, 262
Cephalalgia epileptica, 55
Children's parties, 320
Chloroform, its use in paroxysm, 236
Chorea, 29
CHRISTISON, on uræmia, 116
Church, canons of, 86
CHURCHILL, on puerperal convulsions, 116
Churchmen, their views on demoniacs, 88
Circulation, state of, 77
„ „ in sleep, 150,
„ in brain, 153, 196, 221
„ derangement of, 209
Classification of epilepsy, 190 et seqq.
„ remarks on, 156
Club-foot, its relation to epilepsy, 75
Cod-liver oil, 288
Coitus, 126
Coma, after fit, 9
Comitialis, 10
Confidence of patient, 315
Consciousness in fit, 25
Continence, 142
Contractions of fingers and toes, 5
Convulsions of limbs, 5, 28
„ tonic or clonic, 5
COOPER, Sir A., quoted, 33, 196
COOKE, quoted, 6, 20, 30, 88, 163, 190, 220, 265, 270, 299, 316
COPLAND, quoted, 11, 18, 20, 37, 68, 84, 187, 299
Copper, preparation of, 285
CORNER, on opium, 238

Cotyledon umbilicus, 298
Counter-irritants, 261

DARWIN's case, 159
DAVY's case, 175
Defecation in fit, 8
Definitions of epilepsy, 10
Delirium after fit, 9, 59
Demoniac influences, 85
D'ESPINE, quoted, 114
DELASIAUVE, quoted, 127, 178, 266
Diagnosis, 29
Diathesis, nervous, 63
DICKINSON, quoted, 295
Diet in epilepsy, 121
„ of epileptics, 306
Digestive organs in epilepsy, 121
Digitalis, 294
Diurnal type, 49
DONDERS, on cerebral circulation, 153
Dribbling, 77
Drowsiness after fit, 9
Dry cupping, 250
DURHAM's experiments, 149

EARLE, quoted, 237
Eclampsia, 37
ELLIOTSON, quoted, 107
ENGEL, quoted, 176
Enuresis, 156
Epidemics of epilepsy, 103
Epileptic paroxysm, a part of the disease, 2
Epilepsy, not recognised, 14
„ sequelæ of, 70
„ partial, 30, 65
„ mortality from, 79
„ its universality, 94
„ its frequency, 97
„ a sporadic disease, 102
„ epidemics of, 103
„ its hereditary character, 111
„ its relation to electric discharges, 143

Epilepsy, essential and non-essential, 158
,, pathology of, 161
,, theory of, 189
,, classifications of, 190 et seqq.
,, value of post-mortem researches, 193
,, its relation to other diseases, 213
,, metastasis of, 218
,, state of circulation in, 222
,, acute and chronic, 225
,, cases of, 226, 229
,, medicinal treatment, 223 et seqq.
,, moral influences, 257
,, hygienic treatment, 300 et seqq.
,, proclivity to, 256
ESPARRON, quoted, 65
ESQUIROL, quoted, 15, 26, 65, 69, 78, 107, 178, 186, 254, 257
Exercise, value of, 302
Exorcism, 87
Exostosis, 177
Expression of countenance, 77
Extravasations in fits, 8
Eyes, spasms of, 7

FACCIOLATI, quoted, 10
Face, swelling of, after fit, 9
FARMER, on demoniacs, 88
FARR, quoted, 95, 108
Fear, premonitory, 19
Females more subject to psychical influences, 147
FERRUS, quoted, 167
Fits, frequency of, 42
Fleming, on compression of carotids, 236
Flexors, spasm of, 5
Foramen magnum, contraction of, 177
Formulæ, appendix of, 323
FOVILLE, quoted, 68
FRANK, quoted, 20

FRASER, quoted, 296
Frequency of epilepsy, 97
,, of fits, 42, 62
French statistics, 98
Friedrichshall water, 269
FROMMANN'S case, 284

GALEN, quoted, 10, 15, 35, 44, 126
Galvanism, 239
GEORGET, quoted, 16, 191, 283
GOOLDEN, on sugar in urine, 118
Gheel, colony of, 87
Grand mal, petit mal, 66 et seqq.
GRAVES, quoted, 304
GREGORY, quoted, 18
GUAINERUS, quoted, 316
GUENEAU DE MUSSY's case, 264
GUERIN, quoted, 76
GUGGENBÜHL, quoted, 102
Guinea pigs, epilepsy in, 203

HABIT, influence of, 210
HALFORD, on suppression of urine, 116
HALL, Marshall, quoted, 31, 33, 218, 268
Hallucinations after fit, 59
Headache, 53
,, after fit, 9
Heart, in epilepsy, 121
HÉBRÉARD, quoted, 139
HECKER, quoted, 103
HEMMINGS' Analecta, 231
Hereditary influences, 111
HERPIN, quoted, 113, 133, 138, 275
HIPPOCRATES, quoted, 1, 43, 53, 86
HOFFMANN, quoted, 36, 139
HOLLAND, on somnambulism, 154
HUNT, on epilepsy, 119
,, his treatment of, 286.
Hyperæsthesia in epilepsy, 204

IDELER, on indigo, 297
Idiot, case of epilepsy in, 123
Imbecility after epilepsy, 78

Indigo, 297
Insensibility, 9, 24
Intemperance, 155
Interval, phenomena observed during, 61
Iodide of potassium, 289
Iron, preparations of, 271

Jaw, spasm of, 6
Jones, Handfield, case by, 222
„ remarks by, 224, 231

Kidneys, their relation to epilepsy, 116
Kussmaul and Tenner, quoted, 33, 153, 182, 205, 266

Lallemand, case by, 195
La Mott's case, 160
Laughter, premonitory, 21
Lesions, absence of uniform, 115
Leuret, quoted, 129, 146, 155
Lever, on puerperal convulsions, 117
Life, duration of, in epilepsy, 82
Locock, Sir C., on bromide of potassium, 291
Lotions, their use, 238
Lysons, quoted, 11
„ case by, 241
Lucretius, quoted, 7, 9

Magi, their prescriptions, 316
Malingerer, means of detecting, 4
Mania, dancing, 103
Marriage, 146
Masturbation, 128 et seqq.
Mead, quoted, 43, 88
Measles and convulsions, 120
Measurements, cranial, 168
Medulla oblongata, 179
Memory, 68
Menstruation, 133
Metivié quoted, 187
Melsens, on iodide of potassium, 289
Mental powers, 68, 78

Metastatic epilepsy, 218
Micturition in fit, 8
Mistletoe, 295
Moon, influence of, 43
Moreau, quoted, 19, 46, 52, 76, 107, 110, 139, 146
Morbus comitialis, 10
Morgagni, quoted, 42, 164
Mortality from epilepsy, 79 et seqq.
Muscles, contractions of, 72

Narcotics, 238, 292
Neck, swelling of, 4
„ muscles of, 32
Neuralgia, metastasis of, 218
Newbigging, quoted, 59
Nitro-muriatic acid, 285
Nocturnal type, 49
Nursery, moral management of, 312

Odier, case by, 241
Opiates, 292
Overwork, pernicious influence of, 310
Oxalates in urine, 119

Paget, quoted, 21
Pallor in fit, 4
Paralysis after epilepsy, 70
Parchappe, quoted, 167
Parkes, quoted, 289
Paroxysms, major and minor, 66, et seqq.
„ description of a complete, 4
„ treatment in, 234
Parry, quoted, 35, 65, 236
Pathology of epilepsy, 161 et seqq.
„ value of, 193
Pechlin, quoted, 65
Peiroux, quoted, 18, 20
Penis, erection of, 31
Periodicity, 42, 47
Phosphates in urine, 119
Pineal gland, 170
Pituitary body, 170

PLINY's views on specifics, 315
POUPART, 26
Predisposing causes, 93
Premonitory symptoms, 14 et seqq.
PRESTON, quoted, 266
PRICHARD, quoted, 16, 26, 29, 30, 36, 34, 122, 134, 138, 191, 221, 237, 270
PROUT, on uræmia, 116
Pullna water, 269
Pulse, in the paroxysm, 40
,, in the intervals, 40
Pupil, state of, 7, 64
Purgatives, 267

QUACK remedies, 316

RACE, influence of, 101
RADCLIFFE, on treatment, 220
Ratio of epilepsy, 93
REAGAN, on epilepsy, 219
Reflex, 24
Registrar-General's reports, 95
Relatives of epileptics, 113
Respiration in fit, 8
Rest, its importance, 310
REYNOLDS, Russell, quoted, 32, 191
Riding dangerous, 302
Rodriguez, on indigo, 297
ROMBERG, quoted, 16, 24, 45, 176, 193, 231, 237

ST. DYMPNA's shrine, 87
SALTER, on asthma, 151, 218
SAUVAGES, quoted, 10, 36
Scarlet fever in relation to epilepsy, 120
SCHENCK, quoted, 20
SCHRÖDER VAN DER KOLK, quoted, 39, 125, 143, 179, 198
Scripture, quotations from, 89, 91
Seasons, influence of, 52
Self-control, 219
Seminal discharges in fit, 8
Sequelæ of epilepsy, 70
Setons, 261

Sex, influence of, 106
Sexual organs, 125 et seqq.
SHAKESPEARE, quoted, 86
SHEPPARD, quoted, 317
SHORT's case, 159
Shower-bath, 304
Silver, preparations of, 283
SIMPSON, on puerperal convulsions, 117
SIMS, quoted, 169
Sleep after fit, 58
,, its influence, 48, 149
SMITH, experiments of Dr. Edward, 150
Somnolency after a fit, 58
Spasm, various forms of, 36, 65
,, treatment of, 235
Specifics, 315
Spinal arachnoid, 178
Statistics of epilepsy, 79 et seqq.
,, French, 98 et seqq.
Steel biscuits, 216
Steel cakes for baths, 306
Stimulants, 251
STOPFORD, quoted, 105
Suffocation, sense of, 64
Sugar in urine, 118
Suppers, 309
Symptoms of fit, 4
,, premonitory, 14 et seqq.
Syphilis, its influence, 148

TETANIC spasm, 36
Thunderstorm a cause of epilepsy, 147
Thyroid gland, changes in, 34, 35
TISSOT, quoted, 18, 163, 234, 254, 260, 299, 316
TODD, quoted, 121, 176, 196
Tongue-biting, 38
,, biters, Van der Kolk, on, 181
Trachelismus, 31, et seqq.
,, Kussmaul Tenner on, 208
Training of children, 311
Transition forms, 65 et seqq.
Travelling, value of, 302

TRAVERS, case by, 254
Trephining, 252
TRIPE, quoted, 107
TROUSSEAU, quoted, 293
TULPIUS, quoted, 316
Turpentine, 269
  ,,   Prichard's use of, 122

UREA, excess of, 119
Urine, state of, in epilepsy 116 et seqq.
Uterine functions, 134

VALSALVA'S case, 164
VAN SWIETEN, quoted, 6.
Veins, congestion of, 4

VERRAL'S case, 72
Vertigo, 64, 66, 69
Vitiligo, premonitory, 20
Vomiting in fit, 8
  ,,   premonitory, 20

WATSON, quoted, 12, 18, 107, 128, 205, 270
WEBSTER, quoted, 87
WENZEL'S researches, 169 et seqq., 185
Wine, its use, 308
YEATES, case by, 253

ZINC, preparations of, 275

THE END.

www.ingramcontent.com/pod-product-compliance
Lightning Source LLC
Chambersburg PA
CBHW030308240426
43673CB00040B/1100